More praise for *The Borderless Healthcare Revolution*

"Here in the United States, patients with little access to technology are all too often cut out of healthcare services. Dr. Matt has spent her career working diligently at the intersection of practice and technology to enable access to care and better outcomes. Her stories and research from the field will illuminate and empower a future in which all people, regardless of geography, will have the healthcare access and opportunities they need."

—Betty Rabinowitz MD,
Founder and Former CEO,
EagleDream Health

"This book is not a passive read. It is a challenge. It asks those of us who build, fund, and govern healthcare to stop measuring success by what's technically possible—and start measuring it by whether people can actually get the care they need."

—David Feinberg MD, MBA,
Chairman, Oracle Health and Lifesciences

The Borderless Healthcare Revolution

The Borderless Healthcare Revolution

The Definitive Guide to Breaking Geographic Barriers Through Technology

Sarah Matt, MD, MBA

Foreword by
David Feinberg, MD, MBA,
Chairman of Oracle Health
and Life Sciences

WILEY

Library of Congress Cataloging-in-Publication Data has been applied for:

Hardback ISBN: 9781394357123
ePDF ISBN: 9781394357147
epub ISBN: 9781394357130

Cover Design: Wiley
Cover Image: © ipopba/Getty Images
Author Photo: © InstaHeadshot/aarzoo, inc.
Printed and bound by CPI Group (UK) Ltd, Croydon, CR0 4YY

C9781394357123_151025

For everyone who needed care and couldn't get it, and for the people reimagining what care can be.

Contents

Foreword

Throughout my career, I have had the privilege of leading some of the most innovative healthcare institutions in the world. From UCLA Health to Geisinger, from Google Health to Oracle Cerner, I have worked across systems and sectors with one mission in mind: improving the lives of patients. No matter the setting, no matter the population, one reality has always risen to the top. Access to care remains the greatest unsolved challenge in healthcare.

We have built cutting-edge hospitals. We have trained extraordinary clinicians. We have deployed new technologies with extraordinary potential. Yet still, millions of people go without the care they need. Not because the care does not exist, but because they cannot reach it. They are blocked by geography, by cost, by language, by infrastructure, and by systems that were not built with them in mind. The problem is not capability. It is structure. It is delivery. It is follow-through.

That is what makes *The Borderless Healthcare Revolution* such an important and timely contribution. Dr. Sarah Matt has written a book that does what few others have managed to do. She connects the operational and the human. She moves fluidly from frontline realities to system-level strategy. And she makes a compelling case for how we redesign care not just at the margins, but at the core.

Dr. Matt brings an unusual mix of clinical precision and systems insight. She trained as a surgeon, where the stakes are high and decisions are irreversible. She then stepped boldly into the technology world, helping to lead large-scale transformation efforts in health IT, AI, robotics, and remote surgery. That breadth of experience gives her a perspective that few can match. She understands where care breaks down, but more importantly, she understands how to build pathways that work in the real world.

This book is filled with powerful examples. We meet veterans navigating an impossible maze of benefits and care coordination. We meet migrants who are willing to risk everything to reach a clinic. We meet families in rural America who live closer to broadband towers than to pediatricians. We are reminded that care is not just about diagnosis or procedure. It is about logistics. It is about communication. It is about designing systems that are reachable, not just advanced.

Dr. Matt does not romanticize technology. She is clear about its potential, and equally clear about its limitations. AI can help clinicians triage faster, but it cannot overcome infrastructure that does not exist. Robotics can extend the reach of surgical skill, but only if the network is ready. Digital health can improve outcomes, but only when it is built on trust and delivered with clarity. What sets this book apart is its grounded optimism. Dr. Matt is not waiting for the perfect system. She is working with what is available, identifying where it is already working, and showing how it can grow.

She highlights organizations that are delivering care on street corners, in mobile vans, in rural clinics, and in homes. She walks us through how telehealth can reach communities that traditional health systems often miss. She lays out how remote surgery and AI tools can scale expertise, if thoughtfully deployed. She names the gaps in reimbursement, policy, and training that must be addressed to make these solutions sustainable. And she reminds us that access is not a buzzword. It is a measurable, practical, and actionable goal.

For healthcare executives, this book is a roadmap. For public officials, it is a briefing that connects data with real people. For engineers and technologists, it is a set of use cases with clear requirements and a reality check. And for clinicians like me, it is a powerful reminder of why we entered this field in the first place.

The people we serve are not always coming through our hospital doors. Some live in mountain towns where winter closes the roads. Some live in crowded apartments and cannot take time off work to get to a clinic. Some are caregivers who place everyone else's health ahead of their own. Others are aging alone in places where digital tools could offer the only consistent point of contact. The barriers are diverse, but the message is the same. Unless care is truly reachable, it cannot be considered delivered.

Dr. Matt understands this intuitively. She writes with the urgency of someone who knows the stakes, and with the clarity of someone who has built teams, run programs, and worked through the obstacles.

She is not writing from the sidelines. She is in it. And she is bringing together the perspectives of patients, providers, policymakers, and technologists in a way that few have done before.

The Borderless Healthcare Revolution is not simply a title. It is a directive. It is a shift from care that exists in fixed places to care that exists wherever it is needed. It is a move from systems built for convenience to systems built for connection. It is about reducing the friction, shortening the distance, and delivering care that fits into the lives of real people, not just into facility hours or coverage maps.

I am honored to write this foreword because I believe deeply in what this book stands for. I believe in what Dr. Matt is building. And I believe that when we take seriously the full picture of access; physical, financial, cultural, digital, and informational, we can finally begin to close the gap between what healthcare is and what it should be.

When that happens, we will not just improve systems.

We will restore trust.

We will prevent suffering.

We will fulfill the promise of care that reaches everyone, everywhere.

—David Feinberg, MD, MBA
Chairman Oracle Health and Life Sciences
Former President and CEO, Cerner Corporation
Former Head of Google Health
Former President and CEO, Geisinger
Former President and CEO, UCLA Health

Preface

I didn't write this book because I had time. I wrote it because I couldn't stop seeing the gap between what healthcare *could* be and what it still is.

I've worked in operating rooms, free clinics, and startup war rooms. I've helped build platforms used by millions and walked patients through care plans written in a language they didn't speak. I've been a surgeon, a strategist, a tech exec, and a street medic. Every role taught me something different. The failures all pointed to the same truth.

Access is not only about technology. It's also about design, trust, and execution. The tools matter, but only if they reach the people who need them.

This book is a field guide for people who want to build healthcare differently. For leaders who are tired of hearing, "we can't." For clinicians who know the system doesn't work. For founders and policymakers who want to lead with clarity and urgency instead of buzzwords.

The Borderless Healthcare Revolution is not a manifesto or a blueprint. It's a challenge. If we have the tools, what's stopping us?

You're here because you're part of the answer. Let's get to work.

—Sarah Matt, MD, MBA

About the Author

Sarah Matt, MD, MBA, is a physician, technologist, and systems thinker who has spent her career pushing healthcare to reach the people it too often leaves behind. Trained as a surgeon, she moved early into the technology sector, leading the development and deployment of digital health tools across the globe. Her work has spanned remote robotic surgery on multiple continents, electronic medical record rollouts in under-resourced regions, AI at the Edge, and large-scale patient engagement platforms that transformed access to care.

At Oracle Health, she helped steer enterprise-wide initiatives focused on digital delivery, mobile-first strategies, and clinical cloud infrastructure. She has advised government health programs, collaborated with global health teams, and worked hands-on with startups building platforms for the front lines. Throughout, she has focused not only on what's trendy, but also on what is functional, scalable, and built to last.

Dr. Matt continues to practice medicine, treating patients with complex chronic illness, many of whom are uninsured, unstably housed, or disconnected from traditional healthcare systems. These firsthand experiences with patients who fall outside the mainstream care model continue to inform her approach to innovation and design.

She holds degrees from Cornell University, SUNY Upstate Medical University and the McCombs School of Business at the University of Texas at Austin. She teaches first-year med students at SUNY Upstate and is a frequent speaker at global healthcare and technology forums. Her talks center not on potential, but on execution; how to turn promising tools into meaningful change.

Dr. Matt wrote *The Borderless Healthcare Revolution* to challenge the assumption that access is someone else's problem. Drawing from global case studies, patient interviews, and field experience, she outlines what it takes to deliver care when distance, cost, infrastructure, and trust all stand in the way.

She lives in upstate New York with her husband, Gus, and their four children. Their house is full of curiosity, arguments about dinner, and an unreasonable number of roller skates.

The Geography Problem

From the Front Lines

The summer heat in Texas hits you like a wall when you step outside. On this particular August morning in 2012, I was making my way up the cracked concrete path to Maria's mobile home, my medical bag feeling heavier with each step. The window-mounted air conditioning unit hummed desperately against the rising temperature. This was my third attempt to see Maria; we had previously been unable to find her trailer after two attempts to work through an interpreter, poor phone connections, and constantly changing locations of residence. So here I was, bringing the clinic to her. Maria's story would become one of countless examples I'd encounter of how geography shapes health in ways both obvious and subtle. But let me back up a bit. Let me tell you how I found myself making house calls in the Texas heat, and what it taught me about the deep divide in American healthcare.

When I moved to Austin, TX, I was amazed at the availability of jalapeños. You could literally get them at any restaurant. What's better than jalapeños? Well, tacos for breakfast, lunch, and dinner of course! The city was bustling with music, people, and cutting-edge tech. It was a far cry from Upstate NY where I grew up. At the same time, the stark contrast within the city nearly gave me whiplash.

By night, I navigated the sleek corridors of the UT McCombs MBA program, surrounded by future tech executives and management consultants. These were people of privilege, who were funding advanced degrees, with many of us already managing teams at big firms around town.

But by day, I wound through neighborhoods where the American Dream felt more like a distant mirage, and where few who did not live there would venture. I was providing medical care to patients who couldn't leave their homes. Patients who nobody wanted to see in their offices, because they were so complex, and whose situations were seemingly unfixable. The city's famous breakfast tacos and vibrant music scene had drawn me here, but it was these house calls that would reshape my understanding of American healthcare. Each doorway I stepped through told a different story of how geography, whether measured in miles or mere city blocks, could become an insurmountable barrier to care.

Fact Check

1. In 2022, 7.8 percent of U.S. counties did not have a primary care physician. The national ratio of primary care physicians was 83.8 per 100,000 people.[1]

2. Over the course of the last 10 years, more than 120 rural hospitals have ceased operations, further limiting access to care for populations which are older, less healthy, and less affluent than their urban counterparts.[2]

3. An analysis of 2021 Medicare claims showed that beneficiaries in rural areas received less specialty care than those in urban areas.[3]

---

[1] Bureau of Health Workforce. *State of the Primary Care Workforce Report 2024.* Health Resources and Services Administration (HRSA). Published 2024. Accessed January 17, 2025. https://bhw.hrsa.gov/sites/default/files/bureau-health-workforce/state-of-the-primary-care-workforce-report-2024.pdf.

[2] Chartis Center for Rural Health. *Rural Hospital Closure Crisis Deepens: New Research from Chartis Center for Rural Health Reveals.* Chartis Group. Published 2022. Accessed January 17, 2025. https://www.chartis.com/about/news/rural-hospital-closure-crisis-deepens-new-research-chartis-center-rural-health-reveals.

[3] Association of American Medical Colleges. *Rethinking Rural Health: Issue Brief.* AAMC Research Institute. Published 2023. Accessed January 17, 2025. https://www.aamcresearchinstitute.org/our-work/issue-brief/rethinking-rural-health.

The Reality of Home-Based Care

So, what does "homebound" actually mean? According to Medicare guidelines, homebound patients have conditions that make leaving home difficult without considerable effort and assistance. When I started practicing in the early 2000s, this meant I could see patients monthly; a stark contrast to the typical once- or twice-yearly primary care visits many Americans experience. This increased frequency of care should have meant better outcomes, but the reality was far more complex.

According to the Centers for Medicare & Medicaid Services (CMS), a patient is considered homebound if they meet the following criteria.[4]

Criterion One

- The patient must either:
  - Require the aid of supportive devices such as crutches, canes, wheelchairs, or walkers; the use of special transportation; or the assistance of another person to leave their place of residence due to illness or injury; or
  - Have a condition such that leaving home is medically contraindicated.

Criterion Two

- There must exist a normal inability to leave home; and
- Leaving home must require a considerable and taxing effort.

Even if the patient leaves the home, they may still be considered homebound if the absences are infrequent, of short duration, or for the purpose of receiving healthcare treatment.[4]

Stories from the Field: Physical Barrier Spotlight

The cards were still scattered across Fred's small table when we arrived, evidence of the game we'd interrupted. The air in his apartment was thick with the kind of stillness that comes from windows

---

[4] Centers for Medicare & Medicaid Services. *Medicare Benefit Policy Manual*, Chapter 7, Section 30.1.1. Accessed January 17, 2025. https://www.cms.gov/Regulations and Guidance/Guidance/Manuals/Downloads/bp102 c07.pdf.

rarely opened. "¿Cómo está?" my medical assistant Art asked, his casual Spanish immediately putting Fred at ease. None of us could have known then how the next few hours would unfold, or that this routine visit would become a race against time.

Fred (Frederico per his chart) was a 66-year-old Spanish speaking Latino man living alone in a 65 and up low-income housing apartment on the southeast side of Austin. Two million elderly Medicare patients were completely or mostly homebound in 2011.[5] Fred was just one of this ever-increasing population.

He told me he had "bad feet" for over a decade and scooched along the halls of his apartment complex in a wheelchair at a steady but snail's pace. He was jovial and when we got to his apartment, he had a friend over playing cards. After his wife died almost five years earlier, he had turned to smoking cigars, didn't go to the doctor, and had very few ties with his family. But he had a few friends on his floor in the apartment building and generally seemed to be in a good mood. His social worker had referred him to us. She had helped him get on Medicare, get food stamps, and connect him with a free community service to bring him groceries to his apartment. His home was small and cluttered, not particularly well kept, but then again, he was a bachelor doing his best on his own. These stories from Austin aren't unique to Texas or even to the United States. Across the globe, healthcare systems struggle with similar challenges of access—quality and inclusive care—although their solutions often differ dramatically.

Sometimes I would see new patients without much history to go on, and sometimes it was going to just be a surprise. Less than one-third of primary care physicians in the United States report making home visits, a rate significantly lower than in other high-income countries.[6] This time all we knew was that he had "bad feet." So, when we arrived, we were pretty much ready for anything. My medical assistant fortunately came with me everywhere I went. His Spanish was casual but was extremely helpful when seeing patients. While he had been scolded by more than one abuela for his casual grammar, it was certainly better than my own Spanish, which was abysmal.

---

[5] Ornstein KA, Leff B, Covinsky KE, et al. Epidemiology of the homebound population in the United States. *JAMA Intern Med.* 2015;175(7):1180–1186. Accessed January 17, 2025. https://jamanetwork.com/journals/jamainternal medicine/fullarticle/2296016.

[6] Tikkanen R, Abrams MK. *Finger on the Pulse: Primary Care in the U.S. and Nine Other Countries.* The Commonwealth Fund. Published March 2024. Accessed January 17, 2025. https://www.commonwealthfund.org/publications/issue-briefs/2024/mar/finger-on-pulse-primary-care-us-nine-countries.

We went back and forth with Fred, asking him questions about his living situation, his health history, and any medical problems he might have. He was worried about his feet; they had caused him pain for years and recently they had become even more numb in spots and were "making his socks dirty." But he said he had no other medical issues, was not on any medications, and had not seen a doctor for many years. So, we drew some blood, and I started my exam. He had a low fever to start, his blood pressure was elevated, and his heart rate was increased. But then again, having a doctor in a white coat at their house often stresses folks out a bit. As I moved through my exam, things continued to be rather insignificant.

Then I asked him to remove his socks and shoes for me so I could look at his legs and feet. He said he couldn't do it by himself and admitted that he had been wearing this same set of shoes and socks continuously for over a week straight. We removed the shoes, and a terrible stench filled the room. The socks were stuck onto his feet in multiple places, and after removal, revealed multiple wounds, ice cold toes, and no palpable pulses. But what sealed the deal was the redness spreading up his ankle, and a crackling feeling up his leg with discoloration consistent with necrotizing fasciitis. This is "flesh eating bacteria" for the uninitiated. This is a medical emergency. He had not seen the color and crackling feeling a week earlier when he had gotten help to change his socks. In fact, he hadn't been concerned, because he couldn't feel it. Approximately 5 percent of U.S. adults report forgoing healthcare due to transportation barriers, with higher percentages among Black adults (8%), individuals with low family incomes (14%), and those with public health insurance (12%).[7] Fred couldn't remember the last time he'd been to the doctor, so it was unclear how long his feet had been going downhill.

Once you have seen necrotizing fasciitis (or smelled it!), you never forget it, and this was textbook. It reminded me of cases I had seen during my surgery training. I was convinced of my diagnosis, and this was an emergency. He needed to go to the ER for a surgical consult immediately. I wanted to call 911 and get him seen right away, but he was hesitant. He couldn't pay for an ambulance. He couldn't pay for medications or the copay. He had no one to care for him.

---

[7] Robert Wood Johnson Foundation. *More than One in Five Adults Forgo Healthcare Because of Transportation Barriers.* Published April 2023. Accessed January 17, 2025. https://www.rwjf.org/en/insights/our-research/2023/04/more-than-one-in-five-adults-with-limited-public-transit-access-forgo-healthcare-because-of-transportation-barriers.html.

His feet had been "bad" for years and he had never had an emergent problem. Plus, we just met. How could I help this man?

Fred was one of my most memorable patients, but this is what I saw day in and day out. Much of the time I did more social work than medicine and did what I could to help those I served. But it was backbreaking, emotionally draining, and often straight up saddening when I felt that I didn't have a way to help. These were normal people—parents, sisters, brothers, and grandparents. All of them were suffering in one way or another, yet they just happened to be born on the east side of Austin, without advantages or means, and with no one to advocate for them.

Between patients who struggled with mobility and Fred's hesitation to seek emergency care, a pattern emerged that went beyond simple transportation issues. The barriers to healthcare access form an intricate web of challenges. Consistent issues that presented themselves time and time again revolved around not only transportation, but financial ability and education. Homebound patients, by definition, cannot leave their homes easily, but I was one of very few doctors willing to see them on their own turf in their homes. These patients were seemingly "close" to healthcare but could not actually access it and were in their own personal "healthcare deserts." Transportation to a doctor's office is expensive, could be difficult to arrange, and often was next to impossible to find. Fred didn't have commercial insurance and couldn't pay for copays or anything supplemental for that matter. Lastly, like many in this situation, while Fred knew he had medical issues, he did not fully understand the extent and urgency of them. Trusting healthcare providers in the past had not worked in his favor.

A comprehensive definition of healthcare access incorporates the following interconnected pillars. These are the five pillars of healthcare access (see Diagram 1.1).

1. **Physical Pillar:** This encompasses geographic proximity to healthcare facilities but also includes infrastructure considerations, such as transportation systems and the design of medical facilities to ensure accessibility for disabled individuals.

2. **Financial Pillar:** Healthcare costs often extend beyond medical bills. These include transportation expenses, lost wages, and childcare needs. Out-of-pocket costs can act as significant deterrents to seeking timely care.

3. **Cultural Pillar:** The compatibility of healthcare services with a patient's language, culture, and traditions is critical. Cultural competency builds trust and ensures that patients feel respected and understood.

4. **Digital Pillar:** In the digital age, access to reliable Internet, devices, and digital literacy has become an essential component of healthcare. Patient portals, telehealth visits, and online educational tools depend on these resources.

5. **Trust/Knowledge Pillar:** Historical mistrust due to systemic inequities or past mistreatment can prevent patients from engaging with the healthcare system. Concurrently, knowledge gaps about when and how to seek care can exacerbate health disparities.

Diagram 1-1: Five Pillars of Access.

As you move throughout this text, you'll continue to see where the pillars of access are improved or addressed, as well as where barriers persist or are made worse. Here's an example of how that will look as you move forward:

- **Physical Pillar:** This indicates that the physical pillar is being improved or addressed.
- **Physical Barrier:** This indicates the physical pillar is a barrier or is being made worse.

Observations of Underserved Communities

The map on my office wall told a story, but not the one most people could see. The colorful pins marked hospital locations across central Texas, creating a deceptive impression of comprehensive coverage. What the map didn't show were the stories behind each gap between those pins—the young mother calculating if she can afford the gas to drive three hours for her child's specialist appointment, the elderly farmer who had to move in with his daughter after the local hospital closed, the shift worker who couldn't take time off for medical appointments during regular business hours.

Underserved communities are everywhere, from the most rural, the most remote, to the middle of huge urban centers. Underserved can look very different or may even seem to be very much "served" by the outside world. When we think of rural and remote locations, we often think of places far from the United States, such as Sub-Saharan Africa, the northern provinces of Canada, or isolated islands. But "remote" doesn't necessarily mean thousands of miles away. Remote can mean there is no way to get healthcare within walking distance, within an hour or more, and produces a **Physical Barrier** to care.

Stories from the Field: Physical Barrier Spotlight

One of my kids has a friend at his school whose father I have never met. He's an orthopedic surgeon who works in a small town over an hour away. In speaking to his family, it seems he has a four-year contract and is being compensated quite well. But he is essentially on call 24/7, is the only orthopedic surgeon in an hour radius, and is the only outpatient specialist as well. When the family moved to the beautiful small town in Upstate NY where he is working, they were excited. But they soon realized that the schools were not well funded, and their kids, who were rather advanced, didn't fit in and could not be supported. Now they have an apartment in a bigger city where the mom and three kids live all week. They send the kids to a school that can support them. They visit their actual home and their father on the weekends, which was certainly not their plan when they moved to the area. The father doesn't make it to school events or concerts because he's always on call and over an hour away. This is the double-edged sword of being a rural provider, especially a specialist.

In the United States many counties lack basic services like OB/GYN and cardiology, let alone cancer treatment (see Diagram 1.2). In 2021, about 4.5 billion people, more than half of the global population, were not fully covered by essential health services.[8] Think about the sheer magnitude of that number. That's billions of people worldwide. As for the orthopedic surgeon in my story, his practice was most likely much different than his peers in a larger city. In an area with many orthopedic specialists, you may have one for the hand, one for the spine, one for sports, and others for each subspecialty. However, when you're the only specialist in an area, you'll be taking care of most of those conditions. The concern here is that many local folks won't accept a referral to a specialist farther away since they don't have the means or transportation to go. But from a safety and quality perspective, surgery is a volume practice. Does that surgeon have the volume to safely conduct various procedures with quality for their patients?

---

[8] World Health Organization. *Billions Left Behind on the Path to Universal Health Coverage.* Published September 18, 2023. Accessed January 17, 2025. https://www.who.int/news/item/18-09-2023-billions-left-behind-on-the-path-to-universal-health-coverage.

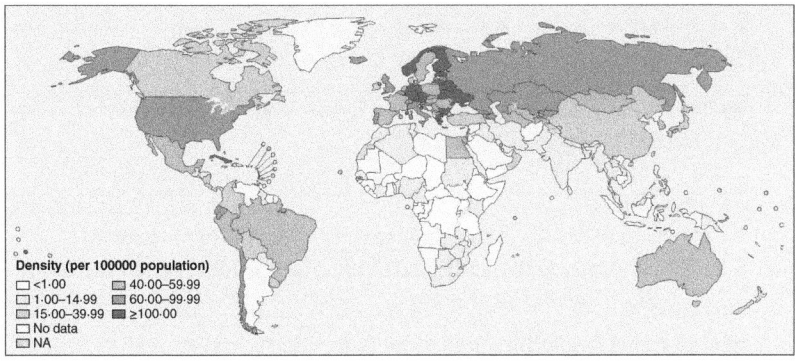

Diagram 1-2: Global Distribution of Surgeons, Anesthesiologists, and Obstetricians per 100,000 Population.[9] The Color of the Map Represents the Density/100,000 People.

It has been shown that patients who do not have access to care have higher rates of chronic diseases like hypertension and diabetes.[10] Usually when they are finally treated, they have progressed to more advanced stages of disease, creating worse outcomes and more costly care. Remember poor Fred and his gangrenous foot. If he had care and intervention much sooner, or better yet preventative care, he may have been in a better spot. Ultimately, I convinced Fred to go to the emergency room, and we called 911 for an ambulance. Working through my assistant for translation was difficult, but we were able to build a level of trust quickly. In his case, he had a below knee amputation of his left leg. He recovered well and received several new services to help him get the support and care he needed, as well as physical therapy, transportation, and help in his home. Fred was in an urban environment but many of the problems remain the same. Rural populations have traditionally lacked effective and frequent preventative care and tend to have higher levels of untreated chronic conditions.

---

[9] Wurdeman T. *Map of Surgeon, Anesthesiologist, and Obstetrician Density*. Program in Global Surgery and Social Change (blog). n.d. https://www.pgssc.org/ workforcemap.

[10] Centers for Disease Control and Prevention. *Chronic Disease Prevalence in the US: Sociodemographic and Geographic Disparities*. Published March 2024. Accessed January 17, 2025. https://www.cdc.gov/pcd/issues/2024/23_0267. htm?utm_source.

Fact Check

- **Emergency Medical Services (EMS) Response Times in the United States:**[11]
  - **Urban Areas:** Average response time is approximately seven minutes.
  - **Rural Areas:** Median response time increases to more than 14 minutes, with nearly 10 percent of encounters experiencing waits of almost 30 minutes.
- **Specialist Availability per 100,000 Residents in the United States:**[12]
  - **Urban Areas:** Approximately 263 specialists per 100,000 residents.
  - **Rural Areas:** Approximately 30 specialists per 100,000 residents.

Reflections on Emotional and Systemic Barriers

Story from the Field: Physical Barrier Spotlight

This past year, Larry, one of the senior members of my volunteer fire department, was diagnosed with a rare cancer. The specialists close by were not equipped to treat him as well as those in a larger city. While his family was doing well and was able to drive him the almost three-hour drive to the large academic cancer center, it was beyond uncomfortable for him. One of our volunteer crews even brought Larry to an appointment via ambulance to see if it would help with his comfort. Nothing helped. Every bump on the drive, no matter how smooth the driver was at the wheel, was excruciating for him. For a man already facing the life and death situation of a cancer diagnosis, this almost put him over the edge of losing all hope and stopping treatment all together. They fortunately found a solution for Larry. His care team was eventually able to get him his infusions locally. Today, he's in remission and surrounded by friends

---

[11] Mell HK, Mumma SN, Hiestand B, Carr BG, Holland T, Stopyra J. Emergency medical services response times in rural, suburban, and urban areas. *JAMA Surg.* 2017;152(10):983–984. doi:10.1001/jamasurg.2017.2230.

[12] Association of American Medical Colleges. *Health Disparities Affect Millions in Rural U.S. Communities.* Published October 31, 2017. Accessed January 17, 2025. https://www.aamc.org/news/health-disparities-affect-millions-rural-us-communities?utm_source.

and family, ready if/when there is a next time. Discomfort and the fear of discomfort almost stopped his care. He was an EMT and well versed in the hospital system, well known in his community, and well connected. But none of these things was able to help him with his issue with access.

While distance (a **Physical Barrier**) and cost constraints (a **Financial Barrier**) are well-known barriers to accessing healthcare, the emotional and systemic hurdles are equally significant and often overlooked. Imagine a member of an underserved population facing not just logistical challenges but also a deep-seated mistrust of the healthcare system. For many minority patients in particular, this distrust stems from historical events, such as the Tuskegee Syphilis Study, where African American men were deceived and denied proper treatment over decades, leading to a legacy of skepticism toward medical institutions.[13] This mistrust compounds psychological **Trust/Knowledge Barriers**, such as fear of discrimination, stigmatization, or feeling unwelcome in healthcare settings. These barriers can deter individuals from seeking care or fully engaging with healthcare providers, even when services are accessible.

Acknowledging and addressing these psychological and historical obstacles is critical to truly improving healthcare access. These are just a few examples of how invisible barriers impact patient outcomes, and I continue to explore them throughout this book. In Larry's case, he was fortunate to have family members with reliable transportation who could drive him. Additionally, he had friends at the firehouse more than willing to drive an out-of-service ambulance at their personal expense. But this is certainly not the norm. There is a significant lack of coordinated transportation services throughout the United States, especially in rural environments. In these areas, catching a bus or a train, or even grabbing a rental car or an Uber can be next to impossible. Instead, these patients have to go to the city center to even catch a ride to the next place. When this means multiple hours of travel just to get to a doctor, it inevitably impacts employment and childcare.

Stories from the Field: Physical Barrier Spotlight

New mothers are experiencing pretty much the hardest job in the world. They have been building a small human, and while they need to heal their own bodies, they also need to care for a helpless infant 24/7.

---

[13] Centers for Disease Control and Prevention. *U.S. Public Health Service Syphilis Study at Tuskegee.* Updated April 22, 2021. Accessed January 18, 2025. https://www.cdc.gov/tuskegee/index.html.

For those without an involved partner, being a single parent makes this 1,000 percent more difficult. Gabby was a 22-year-old African American female seen at a colleague's mother/baby family medicine practice. She had called very concerned that morning about her baby's breathing. He had a bad cough and lots of "snot." She did not have any family members to support her and was raising the child on her own. She had a three-year-old little boy as well, which kept her very busy. When the office got the call, they were happy to make time for them and said she could come right in. She was grateful, and said she would get there as soon as she could.

As the day went on, the staff continued to wait for her in the office, but lunchtime came and went, and she had still not arrived. As the clock hit 4 p.m., the bell rang on the front door, and mother, baby, and big brother all finally arrived. They looked tired; the big brother was complaining about being hungry, and they all looked like they had been in the cold for far too long. When the team finally got them into the room, and of course made sure a lollipop was available for her oldest, she broke down in tears. It turns out she had to walk over a mile with her children to get onto a bus, then transfer at the central bus station. Her son had to use the bathroom, which caused them to miss their transfer, and they finally caught the next bus and had to walk another mile to get to the office. She was exhausted and the kids were cranky. Fortunately, after an exam, it turned out her baby was just fighting off a little cold. She was taught how to use the nose suction to help the baby breathe better and given some tips for getting him to sleep. She was relieved but was not looking forward to the long trip home.

Gabby was a stay-at-home mom, and she was receiving benefits until she could get back to work. If she had been working, think of the time off she would have had to take. Travelling all the way to the office via multiple buses was not just a quick lunch hour appointment; instead she spent the whole day getting to the office. Transportation issues (**Physical Barriers**) can exist in both urban and rural environments. Transportation is considered a systemic barrier to healthcare. While it does impact urban populations, it disproportionately impacts rural populations, creating a cascade of negative consequences for health outcomes and patient-centered healthcare. Rural residents often face significant travel distances to access care due to a shortage of local providers and facilities, with many regions entirely lacking specialists or emergency services. Public transportation options are scarce or nonexistent in these areas, leaving patients to rely on costly or unreliable means of travel, such as personal vehicles or community rideshares, if available at all. This structural challenge exacerbates existing healthcare disparities, as rural individuals are more likely to delay or forgo necessary care altogether, leading to worsening of chronic conditions, preventable complications, and emergency interventions that could have been avoided with routine care. The inability to address transportation needs across settings not only undermines efforts to improve rural healthcare access but also drives up overall healthcare costs by shifting care to emergency and acute settings.[14] Systemic solutions that include investments in transportation infrastructure, mobile health services, and telehealth expansion are critical to addressing this pervasive barrier and ensuring balanced access for rural populations.

I've discussed a multitude of barriers to care that extend from our backyards to the other side of the world. But how can we truly define healthcare access in our modern world? Healthcare access has several components (see Diagram 1.1) and as you can imagine this definition continues to evolve. Think of the differences in communication and technology from just a few decades ago compared to today. Digital "readiness" may not have even been considered a component of this definition at that time, but today more than ever, it is becoming essential. Access to healthcare is more than simply a matter of physical proximity.

The Economics of Rural Healthcare

When examining the **Financial Barrier** aspects of healthcare accessibility, the disparities are striking. Globally, rural communities consistently face a shortage of available medical care. Specialists, such as our friend the orthopedic surgeon, are particularly scarce in these areas. Those who do practice in rural settings often manage extensive workloads, encompassing a broad scope of practice, which may limit their ability to provide optimal care for each patient. So why are there so few rural healthcare services? This shortage is attributed to several factors, including the maldistribution of the healthcare workforce, **Financial Barrier** constraints, and the challenges associated with recruiting and retaining healthcare professionals in rural settings.[15]

---

[14] Rural Health Information Hub. *Transportation to Support Rural Healthcare.* Updated May 2023. Accessed January 18, 2025. https://www.ruralhealthinfo.org/topics/transportation.
[15] Ibid.

Healthcare at its core is a business just like any other industry. A car dealership would not sell cars that were not profitable. A railroad would not keep running train lines where no passengers wanted to ride. Unfortunately, healthcare is in a similar situation. Rural healthcare systems operate in a precarious financial ecosystem, where lower patient volumes and higher operational costs often lead to deficits. In the United States, rural hospitals frequently rely on Medicare and Medicaid reimbursements, which tend to be lower than those from private insurers. Moreover, rural populations often experience higher rates of chronic illness, resulting in increased demand for healthcare services that are costly to provide.

When rural hospitals close, the economic implications are severe (see Diagram 1.3). Between 2010 and 2021, over 136 rural hospitals shut their doors, with 19 closures in 2020 alone, the most in one year for the previous decade.[16] A typical rural hospital supports a significant portion of local employment and contributes substantially to the community's economic stability. The closure of rural hospitals isn't just a medical crisis, it's an economic one. When a rural hospital closes, the impact ripples through the entire community. Beyond the immediate loss of healthcare access, communities lose well-paying jobs, struggle to attract new businesses, and often see property values decline. Rural healthcare systems often serve as the nucleus for ancillary services, including pharmacies, home health agencies, and specialty practices. The loss of these services exacerbates healthcare deserts, forcing patients to travel farther and incur higher out-of-pocket expenses.

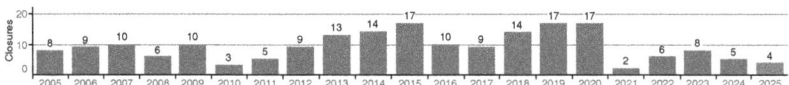

Diagram 1-3: U.S. Rural Hospital Closures 2005–2025.[17]

---

[16] American Hospital Association. *Rural Hospital Closures Threaten Access*. Published September 2022. Accessed January 18, 2025. https://www.aha.org/system/files/media/file/2022/09/rural-hospital-closures-threaten-access-report.pdf?utm_source.

[17] Cecil G. Sheps Center for Health Services Research. *Rural Hospital Closures*. University of North Carolina at Chapel Hill. Accessed January 22, 2025. https://www.shepscenter.unc.edu/programs-projects/rural-health/rural-hospital-closures.

Policy changes and innovative models can mitigate these challenges. Expanding telehealth reimbursement policies, fostering public-private partnerships to sustain rural hospitals, and incentivizing health-care professionals to work in underserved areas are crucial steps forward. Addressing the economics of rural healthcare requires a multifaceted approach. Investments in broadband infrastructure, workforce development, and value-based care models can create more resilient systems. Furthermore, integrating advanced technologies like AI-driven diagnostics and wearable health devices can reduce the burden on overstretched healthcare workers while improving patient outcomes.

The Digital Divide

The role of public libraries in bridging the digital healthcare gap deserves special attention. In many rural communities, libraries serve as de facto telehealth centers, providing both Internet access and private spaces for virtual medical consultations. The American Library Association reports that 98 percent of public libraries provide free Wi-Fi to patrons who otherwise lack access to such resources.[18]

If you're reading this book, and more importantly, if you spent your own money to buy it, then you probably have Internet access at your work or home. Your biggest day-to-day connectivity issue is likely focused on obtaining the highest speed possible to stream your favorite reality show while your kids do their homework simultaneously (or complain that Fortnight is lagging!). However, many people world-wide do not have Internet access as readily as many of us do in the United States. In fact, access to broadband Internet, as opposed to dial-up services, is very limited in many rural areas, even those not far from large urban centers (see Diagram 1.4). Patients in these areas may also lack reliable cell phone coverage, access to broadband or fiber-optic Internet, or may not have Internet access at all. To obtain access, they might need to visit their local library or school during business hours. Globally, the numbers are even more staggering.

---

[18] American Library Association. *Quotable Facts About America's Libraries*. Published 2019. Accessed January 18, 2025. https://www.ala.org/sites/default/files/advocacy/content/ALAquotable%20facts.2019%20web.pdf?utm_source.

As of 2023, approximately 2.6 billion people, or 33 percent of the global population, remained offline.[19]

While cell phone coverage and broadband Internet may not seem like a necessity for quality healthcare, let's think about the patient journey. To make an appointment, how do you contact a doctor's office? To ask a medication or treatment question, do you have to utilize a patient portal or other solution? If you have the benefit of being able to access telemedicine services, do you have the bandwidth to have a clear picture and conversation with your provider? Lacking these means of communication can greatly diminish the ability of a patient (this is even harder for kids) to use these services, creating a **Physical Barrier**. If you're utilizing public resources, having a telemed visit in the middle of the library is certainly not an ideal situation. I have personally taken a telemed visit (as the patient) from the floor of a conference, and I would certainly not recommend this. First, I could barely hear the provider and secondly, I didn't want everyone around me to hear about my personal health issues and history. This was a situation where I had great service and connectivity. If the picture or voice had been choppy, I can't imagine how negative an experience it would have been.

Broadband Access for United States by County

All Races (includes Hispanic/Latino), Both Sexes, All Ages, 2018–2022

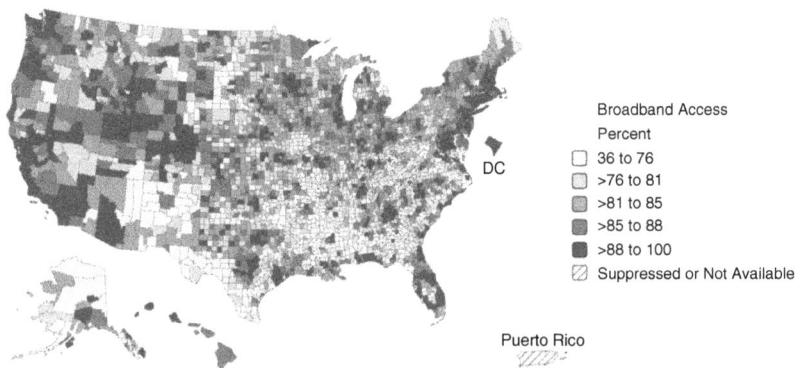

Diagram 1-4: Broadband Access for United States by County (2018–2022 All Races, Both Sexes).[20]

---

[19] International Telecommunication Union. *Facts and Figures 2023 – Internet use*. Published October 10, 2023. Accessed January 18, 2025. https://www.itu.int/itu-d/reports/statistics/2023/10/10/ff23-internet-use.
[20] HD*Pulse*: An Ecosystem of Minority Health and Health Disparities Resources. *National Institute on Minority Health and Health Disparities*. Published February 22, 2025. https://hdpulse.nimhd.nih.gov.

Health Literacy: The Hidden Barrier to Care

Access to broadband and the ability to use it effectively are integral to health literacy. Today, most health information, public health announcements, and patient resources are provided through apps or online links, making printed educational materials increasingly obsolete. Tasks such as reading prescription labels, understanding medical instructions, or navigating patient portals may seem straightforward to healthcare providers, but they present significant challenges for millions worldwide. *Health literacy*—the capacity to access, comprehend, and utilize health information to make informed decisions—is a critical yet often overlooked barrier to healthcare access. The World Health Organization emphasizes that health literacy involves the personal knowledge and competencies that accumulate through daily activities, social interactions, and across generations.[21] These competencies are mediated by organizational structures and the availability of resources that enable individuals to access, understand, appraise, and use information and services to maintain and promote health.

Stories from the Field: Digital Pillar Spotlight

When I was a kid in the age of dial up, my mother had a thick blue book she kept stored in the cabinet right above the family telephone. It contained "all the medical information you could ever need." No matter what illness we had, heard on the news, or could dream up, she had an entry in there to reference. Even as I entered medical school, I couldn't convince her to stop referring to that outdated text. Now many years later she is certainly guilty of occasionally referencing "Dr. Google." Fortunately, she's moved on to more reputable sources from trusted sites and academic medical centers. But if the parent of a physician can potentially have outdated healthcare information presented to them in their day-to-day life, imagine a patient without digital literacy and a pronounced **Digital Barrier**. … I better see if that book is still there!

The challenges of health literacy manifest in countless daily scenarios. Consider a patient with diabetes trying to manage their condition. They must understand blood glucose numbers, calculate insulin doses, interpret nutrition labels, and recognize early warning signs of complications. For someone with limited health literacy, this complex

---

[21] World Health Organization. *Health Literacy*. Published 2023. Accessed January 18, 2025. https://www.who.int/news-room/fact-sheets/detail/health-literacy?utm_source.

self-management can become overwhelming, leading to poor outcomes despite having physical access to care. The digital transformation of healthcare has inadvertently created new literacy challenges. As healthcare systems increasingly rely on patient portals, mobile apps, and telehealth platforms, a new form of literacy, digital health literacy, has become essential. When we think of all the components of health literacy, we can break it down into 10 key components that enable individuals to effectively engage with healthcare information and healthcare services:

1. **Accessing Health Information:** The ability to locate or obtain relevant health data from various sources.

2. **Understanding Health Information:** Comprehending medical terms, instructions, and concepts necessary for informed health decisions.

3. **Evaluating Information:** Assessing the credibility and relevance of health information to one's personal health context.

4. **Communicating Health Information:** Effectively discussing health concerns and information with healthcare providers and support networks.

5. **Navigating Healthcare Systems:** Understanding how to utilize healthcare services, including appointment scheduling and understanding insurance processes.

6. **Decision-Making:** Applying health information to make appropriate health-related choices.

7. **Numeracy Skills:** The everyday skill of making sense of numbers and using basic math. Interpreting and working with numerical health data, such as dosage instructions and medical statistics.

8. **Digital Literacy:** Using digital tools and platforms to access and manage health information.

9. **Cultural Competence:** Recognizing cultural differences that influence health beliefs and practices.

10. **Self-Efficacy:** Having confidence in one's ability to act and make decisions regarding personal health.

From these components, you can see that the diabetic patient just discussed must touch every single one of these areas to properly utilize their insulin and manage their diabetes at home. Any breakdown in even one of these components could lead to an incorrect dose, failure to recognize an emergency, or further complications of disease.

> **Literacy Levers How-To**
>
> Low health literacy drains US $106 billion to $238 billion from the nation's care budget every year; it also drives higher hospital use and mortality in heart-failure patients.[22,23] So what can be done to improve this **Digital Barrier?**
> Three system fixes you can deploy this quarter:
>
> 1. **Adopt Universal Teach Back:** Require every clinician to close each visit with a patient "teach back" confirmation; embed a smart phrase in the EHR (electronic health record) for rapid documentation; audit charts monthly, and target 90 percent completion in six months.
>
> 2. **Launch Multi-Lingual, Plain-Language Portals:** Add an immediate language toggle and eighth grade-reading-level summaries for medications and follow-up tasks. Track portal logins from limited-English-proficiency patients and aim for a 20 percent rise by year end.
>
> 3. **Swap Dense Text for Numeracy Friendly Visuals:** Replace discharge packets and insulin handouts with icon-based dose grids and pictograms. Monitor medication-error reports and seek a 25 percent reduction within 12 months.
>
> Implement these levers together: they attack comprehension, language, and numeracy in one coordinated push, turning the cost burden of low literacy into measurable patient-safety and financial gains.

Rural and underserved communities face compound challenges. In these areas, limited health literacy often intersects with:

- Limited English proficiency
- Lower educational attainment
- Reduced access to technology
- Fewer opportunities for health education

---

[22] World Health Organization. *Health Literacy.* Published 2023. Accessed January 18, 2025. https://www.who.int/news-room/fact-sheets/detail/health-literacy?utm_source.

[23] Vernon JA, Trujillo A, Rosenbaum S, DeBuono B. *Low Health Literacy: Implications for National Health Policy.* Washington, DC: Department of Health Policy, School of Public Health and Health Services, The George Washington University; 2007. Accessed January 18, 2025. https://pmc.ncbi.nlm.nih.gov/articles/PMC7956806.

As you can imagine, cultural context plays a crucial role in health literacy and can become a **Cultural Barrier**. Medical instructions that make perfect sense in one cultural context might be confusing or even offensive in another. For example, dietary recommendations that don't account for cultural food practices or religious restrictions often go unheeded, not because of willful non-compliance, but because they conflict with deeply held cultural values.

The pandemic brought many of these **Cultural Barriers** and **Digital Barriers** into sharp focus. Public health messaging about COVID-19 needed to be understood by diverse populations with varying literacy levels, primary languages, and access. Countries that succeeded in their public health response often employed multi-modal communication strategies: simple visual aids, clear verbal instructions, and culturally appropriate messaging. Those that relied solely on written communication or complex scientific explanations often struggled to achieve public compliance with health measures.[24]

Healthcare providers often overestimate their patients' health literacy. Studies show that even highly educated individuals can struggle with medical terminology and complex healthcare decisions. This "literacy gap" between provider assumptions and patient understanding leads to missed appointments, medication errors, and poor adherence to treatment plans. A study published in the *Journal of Health Communication* found that nurses overestimated patients' health literacy, with overestimates outnumbering underestimates six to one.[25]

The financial implications of poor health literacy extend beyond direct healthcare costs. Lost productivity, increased disability claims, and preventable emergency room visits all stem from inadequate health literacy. For example, heart failure patients with low health literacy are at an increased risk of hospitalization and death.[26] For employers, communities, and healthcare systems, investing in health literacy programs offers a significant return on investment. As healthcare continues to evolve and become more complex, addressing health literacy becomes increasingly critical. The solution isn't simply about

[24] Krieger JL, Neil JM, Duke K, et al. What did the pandemic teach us about effective health communication strategies? *BMC Public Health*. 2022;22:2339. doi:10.1186/s12889-022-14707-3.

[25] Dickens C, Piano MR, Morrow DG, et al. Nurse overestimation of patients' health literacy. *J Health Commun*. 2013;18(Suppl 1):62–69. doi:10.1080/10810730.2013.825670.

[26] Peterson PN, Shetterly SM, Clarke CL, et al. Health literacy and outcomes among patients with heart failure. *JACC Heart Fail*. 2015;3(6):448–456. doi:10.1016/j.jchf.2015.01.014.

making information available; it's about making it accessible, understandable, and actionable to all populations.

Solutions require a multi-faceted approach:

1. Universal precautions in health communication: Assuming all patients may have difficulty understanding health information
2. Integration of health literacy education into school curricula
3. Cultural competency training for healthcare providers
4. Development of user-friendly health technologies

Healthcare's Geographic Divide: A Global Context

Stories from the Field: Physical Barrier Spotlight

Back in my youth, I learned how to scuba dive with a team of other students for a research expedition off a remote Fijian island. A bunch of healthy young people living in a research station among a small, isolated tribe … what could go wrong?

My problems were superficial when you think about it. I got terrible bed bugs and was a walking and talking example of why bed bugs are the absolute worst. I also got double ear infections from diving for data collection multiple times a day. Now an ear infection doesn't seem like a big deal sitting here at my computer all cozy at my home in New York. But it was a much bigger deal on that remote island. We had a limited amount of acetaminophen and other medications with us, and it was for the whole group. I couldn't just go get more at the pharmacy down the street. Secondly, we had to depend on whatever medications were in the provisions we packed weeks before. To this day I still have no idea what I was treated with. The writing on the bottle of ear drops was not in any language I knew, but when you feel that bad, you take the options you've got.

Weeks later when I flew back to the United States, still with a double ear infection, the changes in altitude on the plane were absolutely excruciating. Fortunately, I was able to be treated when I returned home. For those who remained on the island, or for those who grew up there, the ability to get medications or other services meant a two-hour boat ride away. Now looking back, I wonder what we would have done if someone had broken a leg or had a real diving accident that needed urgent evaluation and care? The outcomes would have certainly been different.

Fact Check

■ In 2021, about 4.5 billion people, more than half of the global population, were not fully covered by essential health services.[27]

■ In low-income countries, 50–60 percent of the population lives more than eight kilometers from healthcare facilities, often without access to reliable transportation.[28]

■ In Sub-Saharan Africa, more than 170 million people live over two hours away from a hospital, with 40 percent residing more than four hours away.[29]

Healthcare challenges vary significantly across different regions, often influenced by unique geographical, infrastructural, and socio-economic factors. Here is a high-level overview of just some of these challenges:

Sub-Saharan Africa:[30] Physical and Financial Barriers

■ **Chronic Shortage of Healthcare Workers:** Many countries in Sub-Saharan Africa face a significant deficit in healthcare professionals. For instance, the density of specialist physicians in

---

[27] World Bank. *Billions Left Behind on the Path to Universal Health Coverage.* Published September 18, 2023. Accessed January 18, 2025. https://www.world bank.org/en/news/press-release/2023/09/18/billions-left-behind-on-the-path-to-universal-health-coverage.
[28] Stanford Social Innovation Review. *The Invisible Rural Access Barrier.* Published June 2022. Accessed January 18, 2025. https://ssir.org/articles/entry/the_invisible_rural_access_barrier.
[29] Ballard Brief. *Lack of Access to Maternal Healthcare in Sub-Saharan Africa.* Published December 2021. Accessed January 18, 2025. https://ballardbrief.byu.edu/issue-briefs/lack-of-access-to-maternal-healthcare-in-sub-saharan-africa.
[30] World Health Organization. *Atlas of African Health Statistics 2022: Health Situation Analysis of the WHO African Region.* Published 2022. Accessed January 18, 2025. https://www.afro.who.int/publications/atlas-african-health-statistics-2022-health-situation-analysis-who-african-region-0.

Zimbabwe is considerably lower than the global average, with fewer than 1 physician per 10,000 individuals.

- **Lack of Basic Infrastructure:** Healthcare facilities often struggle with inadequate infrastructure, including limited access to essential services like electricity and clean water. This hampers the delivery of quality healthcare services.

- **Limited Access to Essential Medicines:** The availability of necessary medications is often restricted due to supply chain issues and financial constraints affecting patient care.

South Asia: Physical, Financial, and Cultural Barriers

- **Overwhelming Population Density Straining Limited Resources:** High population densities in countries like India and Bangladesh place immense pressure on existing healthcare systems, leading to resource constraints.

- **Geographic Barriers:** Regions with challenging terrains, such as the Himalayas, impede access to healthcare services, making it difficult for populations to receive timely medical attention.

- **Cultural and Gender-Based Barriers to Care:** Societal norms and gender roles can limit access to healthcare for certain groups, particularly women, affecting overall health outcomes.

Remote Island Nations: Physical and Financial Barriers

- **Extreme Isolation from Specialized Care:** Isolated locations result in limited access to specialized medical services, necessitating travel to distant facilities for advanced treatments.

- **Vulnerability to Natural Disasters Disrupting Medical Supply Chains:** Natural calamities like cyclones and tsunamis can sever supply lines, leading to shortages of essential medical supplies.

- **Limited Capacity for Emergency Medical Transport:** Geographical isolation and limited infrastructure hinder the ability to provide prompt emergency medical services.

Disparity in Emergency Response Capabilities

- **Urban Singapore:** Boasts an average emergency response time of approximately eight minutes, reflecting a well-developed infrastructure.[31]
- **Urban India:** Mean response times vary from 14 to 47 minutes on average, highlighting significant gaps in emergency services.[32]
- **Remote African Villages:** Often lack formal emergency response systems, leaving communities without timely medical assistance.

These disparities underscore the critical need for tailored healthcare strategies that address the unique challenges of each region, focusing on infrastructure development, workforce enhancement, and balanced resource distribution. But the **Physical Barriers** in global healthcare don't exist in isolation; they are intensified by an interconnected web of modern challenges. Climate change has emerged as a particularly insidious force in healthcare delivery. As extreme weather events increase in frequency and severity, they don't just disrupt existing healthcare services; they fundamentally reshape the landscape of healthcare access. In Bangladesh, for example, rising sea levels and increasingly severe monsoons regularly cut off coastal communities from medical care for weeks at a time. These communities face a cruel double burden: their healthcare needs increase due to climate-related illnesses while their ability to access care diminishes.

Political instability can also significantly complicate healthcare delivery. In conflict zones, challenges extend beyond **Physical Barriers** to include navigating dangerous areas where healthcare infrastructure has been deliberately or inadvertently destroyed. Healthcare workers in these regions often flee, leading to a devastating "brain drain" that can take generations to recover from. For example, in Syria, the emigration factor of physicians to the United States is around

---

[31] Ministry of Home Affairs, Singapore. *Oral Reply to Parliamentary Question on Average Response Times for Emergency Vehicles to Reach Their Destinations in the Past Three Years.* Published March 21, 2023. Accessed January 20, 2025. https://www.mha.gov.sg/mediaroom/parliamentary/oral-reply-to-pq-on-average-response-times-for-emergency-vehicles-to-reach-their-destinations-in-the-past-three-years.

[32] Factly. *What is the State of Emergency Ambulance Services in India?* Published January 2023. Accessed January 20, 2025. https://factly.in/what-is-the-state-of-emergency-ambulance-services-in-india.

13 percent, indicating a substantial loss of medical professionals.[33] The ripple effects extend beyond the conflict zones themselves, as neighboring countries struggle to absorb and care for refugee populations, often in makeshift facilities with limited resources. There are over a billion people currently living in settings of conflict, displacement, and natural disasters, which severely hampers the delivery of basic health services.

Economic disparities perhaps create the most persistent issue— the **Financial Barrier** to improving global healthcare access. While wealthy nations debate the latest advances in telemedicine and robotic surgery, many developing countries struggle to maintain basic medical supplies and infrastructure. The cruel irony is that the region's most in need of innovative healthcare solutions are often the least able to afford them. A single medical helicopter for rural emergency response, a common sight in developed nations, could consume the entire annual healthcare budget of a small developing country. This economic reality forces difficult choices: does a nation invest in basic primary care centers that might help many, or emergency transport systems that could save fewer but critically ill patients? This complex interplay of challenges requires solutions that go beyond simply building more clinics or training more doctors. Any effective approach to global healthcare access must consider these compounding factors and their long-term implications for healthcare delivery.

The COVID-19 Effect

One of the areas that increased at a rapid pace during the pandemic was telemedicine. All elective surgeries at this time were on hold, and unless you had an emergent situation, you did not want to be in a doctor's office or a hospital emergency room. If you went, you were almost guaranteed to get infected. My family's first telemedicine appointment was in the summer of 2020. My son needed a medication refill, and his doctor, who would normally require an in-person visit, had to adapt. It was nothing fancy: just a phone call. But it was effective. She gathered the information she needed and called in the refill. This simple interaction exemplified how quickly medicine had to transform.

---

[33] Thieme Connect. *Migration of Physicians and Health Care Workers: Syria as an Example.* Published 2012. Accessed January 20, 2025. https://www.thieme-connect.com/products/ejournals/pdf/10.4103/2231-0770.94802.pdf.

As the pandemic continued, there was also a shift in the expectations of providers and patients, and this has continued even today:

- Patients became more engaged in their care decisions.
- Healthcare consumers became more discerning, especially in high income countries.
- Technology adoption accelerated across age groups.
- Traditional care models were questioned and reimagined in all settings.

Fact Check

- U.S. telehealth visits surged by 154 percent in late March 2020 compared to the same period in 2019.[34]
- A full 84 percent of physicians were offering virtual visits by April 2021, and 5 percent preferred to continue those services in the United States.[35]
- Medicare telehealth visits increased from approximately 840,000 in 2019 to 52.7 million in 2020.[36]

The rapid adoption of telehealth during COVID-19 catalyzed a broader revolution in healthcare delivery. What began as a stopgap measure eventually evolved into a fundamental restructuring of how we think about care access. Early in the pandemic, I witnessed my son's pediatrician transform her practice overnight from waiting rooms and in-person visits, to a hybrid model of care delivery. By the end of 2020, her practice had not only mastered telehealth but had implemented remote monitoring for their young asthma patients. This transformation

---

[34] Centers for Disease Control and Prevention. *Trends in the Use of Telehealth During the Emergence of the COVID-19 Pandemic — United States, January–March 2020.* Published October 30, 2020. Accessed January 20, 2025. https://stacks.cdc.gov/view/cdc/97233.

[35] American Hospital Association. *A Fresh Perspective: Where Telehealth Growth Will Go From Here.* Published July 20, 2021. Accessed January 20, 2025. https://www.aha.org/aha-center-health-innovation-market-scan/2021-07-20-fresh-perspective-where-telehealth-growth-will.

[36] U.S. Department of Health and Human Services. *Medicare Beneficiaries' Use of Telehealth Services: A Closer Look.* Published December 2021. Accessed January 20, 2025. https://aspe.hhs.gov/sites/default/files/documents/a1d5d810fe3433e18b192be42dbf2351/medicare-telehealth-report.pdf.

occurred across multiple fronts, including different disease processes, settings of care, and in populations not previously served.

Perhaps one of the most significant innovations was the rapid development of virtual specialty networks. Rural hospitals that previously struggled to access specialist consultations could suddenly connect with urban medical centers for real-time consultations. However, this wave of innovation, while impressive, exposed critical gaps in our healthcare infrastructure. The digital divide, long a concern in healthcare delivery, became a chasm that threatened to leave vulnerable populations even further behind.

The financial landscape of healthcare underwent its own revolution during this period. Payment models that had remained relatively static for decades suddenly required rapid evolution to accommodate new care delivery methods. Insurance companies and Medicare scrambled to develop telehealth reimbursement structures that made sense for providers and patients. Remote patient monitoring, previously a niche service, needed new payment codes to reflect its growing importance. States rushed to implement virtual care parity laws, ensuring that providers would be compensated fairly for virtual visits.

As we look to the future, the innovations sparked by COVID-19 continue to evolve in exciting ways. Healthcare delivery is increasingly moving toward a hybrid model that combines the best aspects of traditional and virtual care. Home-based care, supported by remote monitoring technologies, is becoming more sophisticated and widely available. Platform-based healthcare solutions are making it easier to coordinate care across multiple providers and settings. The key lesson from this period of rapid innovation isn't just about technology adoption, it's about adaptability and inclusive care. As one of my colleagues noted, "The technology was always there. What changed was our willingness to use it and our understanding of who might be left behind." This understanding must guide our path forward as we continue to innovate in healthcare delivery.

A New Map

When I first arrived at Fred's apartment, I didn't know we'd be calling 911. I didn't know he would lose his leg or that he'd later tell me it saved his life. What I did know, from the moment I stepped through his door, was that we needed a better map.

Not a map of hospitals or clinics—those already existed. We needed a map that reflected the actual terrain of healthcare access; one that accounted for distance and dollars, culture and trust, digital reach and systemic red tape. A map that understood how someone like Fred could live just minutes from world-class healthcare and still be medically stranded.

This book is that map.

Over the next chapters, I walk through the **Five Pillars of Access—Physical, Financial, Cultural, Digital,** and **Trust/Knowledge**—and examine how barriers to this access appear in cities, towns, and villages across the globe. I explore how emerging technologies like telehealth, artificial intelligence, and remote robotic surgery are expanding care in places that were once unreachable. But I'll also ask hard questions: When does technology widen the divide instead of closing it? What does it take to build systems that actually work for the people left behind?

Fred's story is not just a cautionary tale. It's a call for revolution. Because access isn't a single doorway; it's a network. A living, shifting, interdependent web of people, policies, platforms, and pain points. If we want to create a future where everyone gets the care they need, we have to learn to see that web and start pulling the right threads.

Let's begin.

 Startup Builder's Box

You're not building for "rural" or "urban." You're building for real people who miss care because the clinic is one transfer too far, the sidewalks aren't plowed, or the copay isn't worth the risk. Want to design for access? Start here:

1. **Don't Assume the Infrastructure Exists:** Most care plans assume patients have reliable transport, a working phone, or broadband. They often don't. Consider if your segment needs you to build for low-bandwidth, no-car, no-Wi-Fi, and walk-only environments. If your solution needs GPS, cellular signal, or real-time video, ask what happens when none are available.

2. **Go to Them:** Your target user may live three miles from a major hospital and still be medically unreachable. Bring care to the bus stop, the food pantry, or the church basement. Think about mobile units, microclinics, or asynchronous care that fits inside a pharmacy kiosk.

3. **Solve for Friction, Not Distance:** Fred's story wasn't about miles; it was about socks, silence, and no one showing up. Treat care deserts like logistics problems. Where does care fail to arrive? What can you hand off to community partners, volunteers, or automation? Where does your design quietly assume privilege?

If your product doesn't close the gap between *nearby* and *reachable*, you're not solving for access. You're just mapping hospitals.

Leaders' End-of-Chapter Action Checklist: Chapter 1: "The Geography Problem"

LEADER	HIGH-IMPACT ACTION TO STRENGTHEN *PHYSICAL ACCESS*
❑ Board Director	Approve a mobile-clinic capital line and mandate a quarterly *Travel-Time Index*, aiming for <30 minutes to primary care for 90% of residents.
❑ Chief Executive Officer	Charter a *Road-to-Care Task Force* and cut average patient mileage reimbursement claims by 25% within 12 months.
❑ Chief Information Officer	Publish a county broadband gap map on the intranet and broker ISP partnerships that raise household telehealth capability to 95%.
❑ Chief Health Information Officer	Standardize low-bandwidth tele-triage protocols and reduce avoidable rural transfers to tertiary centers by 15% over the next year.
❑ VP Clinical Operations	Rotate "pop-up" specialty clinics through every critical-access site each Friday and track visit volume plus no-show rates.
❑ VP Nursing & Patient Education	Launch rideshare or fuel-card vouchers for high-risk patients; target a 20% drop in missed appointments after six months.
❑ VP Data & Analytics	Build a geospatial dashboard overlaying service lines with the Social Vulnerability Index to select two priority ZIP codes each quarter.
❑ Telehealth Program Manager	Field-test 10 low-bandwidth telehealth kits monthly in homebound households and report latency plus clinical completion rates.
❑ Patient Experience Manager	Add a one-click transit-burden question to every post-visit survey and include results in the weekly *Voice of the Patient* digest.
❑ Community Health Worker Supervisor	Host "Bus-Stop Clinics" for blood-pressure checks and schedule twice weekly; enroll 100 new patients by year-end.
❑ Director of Snacks & Morale	Procure GPS-enabled donuts so nobody gets lost on the way to care; distribute at the next quality huddle with extra sprinkles for punctuality.

The Digital Revolution in Healthcare

A Shift in My Perspective

My first realization that I was in the midst of this digital revolution was when my kid's pediatrician's office introduced a patient portal. Now, I know patient portals don't have the best reputation, but as a parent of a gaggle of kids, it felt revolutionary at the time. Every year, I faced the same ordeal: scrambling to get my children's immunization records in time for school. I'd call the pediatrician's office, beg for the records, and then anxiously wait for them to arrive via snail mail, often holding my breath, hoping they'd make it before the school nurse's deadline. On more than one occasion, delays and errors caused chaos, especially since I have a different last name than my kids. When we lived near state borders, I had to juggle care across two states. One time I recall wishing I had my own fax machine at home so they could send it to me directly. Wishing for a fax machine, now that's desperation!

Then came the portal. For the first time, I could access the records instantly, on my own. No more phone calls, no more waiting, no more anx-

Wishing for a fax machine, now that's desperation!

iety. What had once felt like a minor miracle was, in hindsight, a glimpse of how technology could empower patients and transform the experience of care. It wasn't just convenient; it made me feel like

I had some control over a chaotic system. It wasn't perfect by any means, but it decreased the pain and suffering of school forms for this mom!

Stories from the Field: Digital Pillar Spotlight

The second turning point came when I got my first Fitbit. This was back when the model was nothing more than a small pod you clipped into a wristband or waistband to count steps. That's basically all it did, but I was hooked. I'd hold "walking meetings" with my direct reports and even get in steps during lunch. Something about having a tiny device that tracked even one aspect of my health and wellness felt like a window into the future. It made me wonder: if this technology could count steps, what else could it measure? What could it mean for monitoring patients when they weren't in the doctor's office or hospital? It was a small device, but it opened a world of possibilities in my mind. For the first time, I saw how technology could extend the reach of care beyond traditional boundaries, turning everyday tools into potential lifesavers.

These experiences, one deeply personal and one technological, began to chip away at my skepticism about digital health. The portal

Diagram 2-1: Five Pillars Snapshot.

showed me how technology could simplify and streamline access to critical information, while the Fitbit hinted at the transformative potential of remote monitoring. Together, they made me believe that digital health wasn't just a buzzword; it was the next wave of meaningful change in healthcare (see Diagram 2.1).

My perspective began to further shift during my tenure at Oracle Health. There, I confronted the sheer scale of inefficiencies inherent in traditional healthcare systems and at the global level. For the first time, I saw how EHRs (electronic health records) could streamline workflows and improve access to patient data at hyperscale. But it wasn't until we acquired Cerner (the biggest EHR in the world at the time) that I fully recognized the transformative power of digital health. There was so much opportunity to improve what was already there, but at the same time it was like fixing the plane while you are flying it. Everything was critical, there was no room for error, and the need for patient safety was constant.

Fact Check

1. U.S. national health spending is estimated to grow at an average annual rate of 5.4 percent from 2019 to 2028, eventually reaching $6.2 trillion by 2028.[1]

2. The global digital health market is projected to reach $81 billion by 2030, growing at a compound annual growth rate (CAGR) of 19.5 percent from 2024 to 2030. This includes categories such as telehealth, health IT, mobile health apps, wearable devices, and personalized medicine platforms.[2]

3. As of 2024, approximately 535 million wearable devices were in use globally—including fitness trackers, smartwatches, hearables, and smart clothing—with the market growing at about 5.4 percent year-over-year.[3]

---

[1] U+ Insight. *The State of Digital Transformation in Healthcare.* Accessed January 28, 2025. https://www.u.plus/insights/the-state-of-digital-transformation-in-healthcare.

[2] Grand View Research. *U.S. Digital Health Market Size, Share & Trends Analysis Report by Technology, by Component, by Region, and Segment Forecasts, 2024–2030.* Published 2024. Accessed July 23, 2025. https://www.grandviewresearch.com/industry-analysis/us-digital-health-market-report.

[3] International Data Corporation (IDC). *Wearables Market Forecast and Tracker.* Accessed July 23, 2025. https://www.idc.com/promo/wearable vendor/.

The Rise of Electronic Health Records (EHRs) and Telehealth

In the early 2000s a digital revolution was unfolding inside hospitals and clinics: the rise of EHRs and telehealth. These two innovations were happening in parallel: one solving the problem of distance, the other aiming to improve data availability. Yet, unlike telehealth, EHR adoption wasn't always met with enthusiasm. The adoption of EHRs marked a pivotal step in the journey toward digital healthcare and would make telehealth possible. The transition from paper-based records to EHRs promised a revolution in healthcare efficiency and data management. EHRs promised improved efficiency, streamlined workflows, and better patient outcomes, but implementations were often fraught with challenges.

Early implementation efforts revealed numerous technical, operational, and cultural challenges that impacted healthcare providers and systems across the globe. Resistance from providers, disruptions to existing workflows, and the sheer complexity of overhauling traditional systems meant that successful EHR implementations were rare. Understanding these obstacles sheds light on the evolution of EHRs and the journey toward the future of digital healthcare.

This handful of obstacles feels logical, but in the complexities of a large-scale implementation, it always seems like someone or something gets missed. Let's dig into each obstacle and consider more deeply how these obstacles can get in the way, and perhaps where they have been addressed.

High Implementation Costs

The high cost of an EHR implementation has been a **Financial Barrier** from the start, particularly for smaller practices and rural hospitals. Beyond the initial investment in software licensing and hardware upgrades, organizations often underestimate the ongoing financial burden, including maintenance, data migration, system downtime, staff training, and more. Lost productivity during the transition further adds to expenses. Given that total costs frequently exceed initial estimates, long-term financial planning is essential (see Diagram 2.2).

Initial Implementation Costs:

- **Software & Licensing:** One-time purchase or subscription fees
- **Hardware:** Servers, computers, tablets, and printers
- **Installation & Customization:** System setup and workflow integration
- **Training:** Staff education on EHR use
- **Data Migration:** Transferring existing patient records
- **Compliance & Security:** HIPAA and cybersecurity measures

Ongoing Maintenance Costs:

- **Software Updates & Licensing:** Regular upgrades and compliance adjustments
- **IT Support:** Technical assistance and troubleshooting
- **Hosting & Security:** Cloud-based fees or on-premises server maintenance
- **Training & Retraining:** Ongoing education for staff
- **Regulatory Compliance:** Adapting to evolving healthcare laws and policies

Indirect Costs:

- **Productivity Losses:** Temporary inefficiencies during transition
- **Workflow Adjustments:** Changes in practice operations
- **Data Backup & Recovery:** Ensuring data integrity and security
- **Legal & Compliance Risks:** Potential fines or breaches

Cost Estimates:

- **Small practices:** ~$30,000–$50,000 per provider (initial), ~$4000–$10,000 per year (maintenance)
- **Large organizations:** Costs in the millions, depending on scale

Several studies estimate the cost of purchasing and installing an electronic health record (EHR) ranges from $15,000 to $70,000 per provider in the outpatient setting.

Diagram 2-2: General Components of Cost of EHRs for Provider Practices in the United States.[4]

Resistance from Clinicians and Staff

Early EHR systems were often met with skepticism and outright resistance from healthcare providers. *Let's not pretend, they are still a major hassle!* Many clinicians felt these systems disrupted workflows, adding administrative burdens while detracting from direct patient care. Poorly designed interfaces and complex documentation

---

[4] Office of the National Coordinator for Health Information Technology. *How Much Is It Going to Cost Me?* HealthIT.gov. Accessed January 28, 2025. https://www.healthit.gov/faq/how-much-going-cost-me.

requirements exacerbated frustrations, leading to inefficiencies rather than the promised improvements. Instead of streamlining care, EHRs often created "click fatigue," requiring excessive time spent navigating systems rather than engaging with patients. Unfortunately, these same **Digital Barrier** challenges persist in digital health transformations today.

Stories from the Field: Digital Barrier Spotlight

Throughout my career, I have balanced clinical work with healthcare technology leadership, always striving to keep patient care as part of my edict. To stay grounded, I have worked at urgent care centers and charity clinics along the way. My first urgent care experience was purely practical: I needed flexible hours while leading product development at a healthcare software company. The small practice, run by a father and two sons, handled mostly workers' compensation cases, DOT physicals for truck drivers, and entrance exams for new sheriff's department recruits. Each day brought a mix of minor injuries, government paperwork, and the occasional adventure: like extracting a gazillion bee stingers from a patient who had accidentally mowed over a nest, twice!

The practice was later acquired by a larger urgent care network, forcing a transition from the homegrown EHR to a standardized corporate system. Since I worked only one shift per week, my training consisted of a quick 30-minute tutorial from the practice manager before being thrown into a fully new system. Surprisingly, this was more training than I had received during medical school or residency when encountering new EHRs! I adapted quickly, but many colleagues struggled.

Less than a year later, another acquisition introduced even greater changes. This time, the restructuring was severe: nurses were eliminated from the workflow entirely, leaving only medical assistants to support day-to-day patient care. Suddenly, tasks I had relied on the nursing team to perform were now my responsibility. With no warning, I found myself frantically Googling "how to give a tetanus shot" while searching the medication fridge for supplies. And of course, with the acquisition came another EHR transition, two system overhauls in less than a year. I adjusted, but not everyone could. One senior physician, unable to keep up with the constant changes, ultimately retired rather than endure yet another learning curve. His story is far from unique. Many providers, especially those who trained before widespread EHR adoption, struggled to adapt. Some choose early retirement over learning new workflows, while others experienced burnout from inefficient systems that were never designed with their needs in mind.

The resistance clinicians feel toward EHRs (even today!) is not due to an inherent aversion to technology but rather the way these systems have been implemented. Several key factors contribute to this ongoing struggle:

- **Inefficient Workflows:** Early EHRs were built as digital filing cabinets centered on billing, rather than tools for clinical efficiency. Even today, many systems require excessive clicks for simple tasks and hours of documentation even after clinic hours.

- **Lack of Leadership:** In the early days of EHR adoption, decisions were often made by the most tech-savvy physician available, sometimes leading to DIY system development. Without structured leadership roles—such as Chief Medical Information Officers (CMIOs)/Chief Nursing Information Officers (CNIOs)/ and the more modern role, Chief Health information Officers (CHIOs)—technology decisions were mostly disconnected from clinical realities.

- **Failure to Prioritize Usability:** Since EHRs were designed with billing and administrative priorities in mind rather than physician workflows, providers often spend more time completing documentation than engaging in direct patient care.

- **Disruptions to Patient Interaction:** Many studies assessing EHR implementation have found mixed results. While some data suggests efficiency gains, providers often *feel* like they are spending less time with their patients. Patients too often complain that their provider is typing or "on the computer" during their visit instead of making actual eye contact with them. Regardless of the research, you will rarely meet a doctor who truly loves their EHR.

- **Training Deficiencies:** Proper training is one of the strongest predictors of EHR satisfaction, yet it is often neglected or rushed.

The potential of EHRs to improve care coordination, reduce errors, and empower patients has driven adoption despite these challenges (see Diagram 2.3). However, successful implementation requires more than just rolling out new software. Some of the most effective transitions have prioritized *change*

> The only thing providers hate more than what they have is something new.

management: engaging frontline staff early, gathering feedback, and providing strong leadership throughout the process.

Stories from the Field: Trust/Knowledge Pillar Spotlight

Reflecting on my experiences, I've rarely witnessed a truly "smooth" EHR implementation; but I have seen projects where thoughtful planning and strong leadership mitigated the inevitable difficulties. One of the most impressive efforts I've observed was Oracle Health's phased rollout at AtlantiCare beginning in 2024.[5]

This project was ambitious, aiming to integrate multiple solutions into the hospital system over several years. While challenges were inevitable, one particularly successful aspect was the introduction of the Oracle Digital Assistant. This tool was first deployed to 50 providers in outpatient offices, with specialties well-suited for its capabilities.[6] The implementation team wisely selected a group of provider champions: enthusiastic clinicians who supported the effort and provided invaluable feedback during the pilot phase.

What set AtlantiCare's approach apart was its emphasis on proper change management. Recognizing that technical hurdles and workflow adjustments were unavoidable, the leadership team prioritized staff engagement. They invited every employee in the system to tour a special exhibit showcasing the hospital's history alongside live demos of the new solutions. By involving the entire organization and fostering a sense of ownership, AtlantiCare ensured that staff felt heard and empowered during the transition. "Part of seeing that increase in our adoption rate has been that [Oracle is] listening to the feedback," reported Jordan Ruch, CIO of AtlantiCare, in an interview with Beckers. "They're making enhancements to the system. The providers see that, and they're excited to help shape the creation of the tool." This focus on communication and inclusion was reinforced by strong support from leadership, which cascaded down through the organization. Ultimately, the success of this project wasn't just about the technology, it was about creating a culture where staff felt informed and invested in the outcome.

---

[5] AtlantiCare announces Vision 2030. *AtlantiCare*. Accessed January 28, 2025. https://www.atlanticare.org/news/atlanticare-announces-vision-2030.

[6] AtlantiCare sees 80% adoption of Oracle's AI agent. *Becker's Hospital Review*. Published January 2025. Accessed January 28, 2025. https://www.beckershospitalreview.com/digital-health/atlanticare-sees-80-adoption-of-oracles-ai-agent.html.

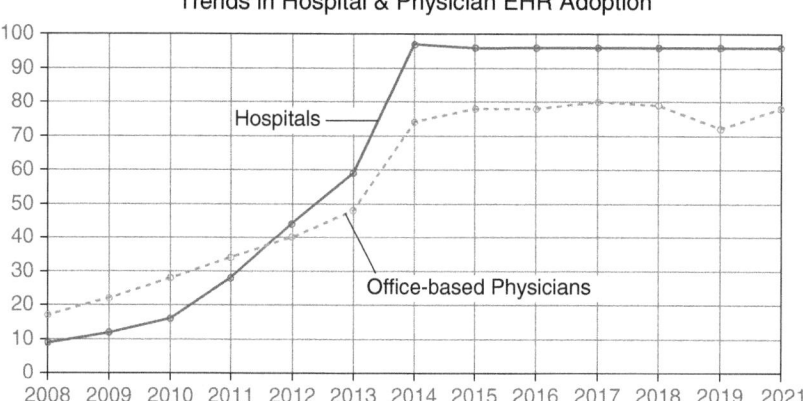

Diagram 2-3: EHR Adoption over Time; Percentage of U.S. Healthcare Providers and Hospitals Adopting EHRs from 2008 to 2021.[7]

EHRs have come a long way, but challenges remain. Physicians dislike change, but they dislike inefficient systems even more. If EHRs, or any healthcare technology adoption is to succeed, platforms must evolve to prioritize usability, integrate seamlessly into workflows, and provide genuine support rather than additional burdens. The key to reducing resistance is simple: listen to clinicians, design for usability, and support providers throughout the transition.

Fact Check

1. U.S. hospitals with advanced digital maturity reported significantly stronger patient experience outcomes, particularly in areas such as communication with nurses and doctors, as well as communication about medications and therapies.[8]

2. More than 50 percent of EHR systems either fail or fail to be properly utilized.

(continues)

[7] Office of the National Coordinator for Health Information Technology. *National Trends in Hospital and Physician Adoption of Electronic Health Records*. Published 2023. Accessed January 27, 2025. https://www.healthit.gov/data/quickstats/national-trends-hospital-and-physician-adoption-electronic-health-records.

[8] Smith JD, Johnson MK, Chen Y. Evaluating the impact of digital maturity on patient experience in U.S. hospitals. *J Health Inform Res*. 2022;14(3):567–589. doi:10.1007/s41666-022-00099-3.

(continued)

3. A staggering 74 percent of clinicians noted increased work hours post-EHR implementation, with 71 percent attributing burnout to EHRs.[9]

Interoperability Matters

One of the most glaring issues with EHRs has been their inability to communicate with other systems. Vendors originally developed proprietary platforms with little regard for data-sharing standards, resulting in silos of information that can hinder care coordination. This lack of interoperability creates significant challenges in transferring patient records between providers or across health systems.

Stories from the Field: Digital Barrier Spotlight

In the early 2000s, digital health was an emerging frontier—a blend of innovation, trial-and-error, and skepticism. As a patient, I still relied on phone calls to contact my doctor's office; a process often riddled with dropped calls, long waits, and information from the office that only came in the form of printed handouts. As a surgical resident, I experienced the patchwork of digital health systems firsthand, and my perspective was firmly rooted in skepticism.

At Washington Hospital Center, where I completed my primary training, we had a hybrid system—paper notes in charts, triplicate forms for consults, and lab results accessible through a basic computer system. The Emergency Department, however, had almost fully digitized its operations. Rotating through other hospitals, I encountered an eclectic mix of systems; some entirely paper-based, some leaned on homegrown software, while others were miles ahead on an enterprise solution. Training was nonexistent; I was often handed a username and password by a fellow resident and given five minutes of rushed instructions before diving into a new system. When I say I was handed a username and password, I literally mean they handed me "their" username and password! My fear of harming patients and disappointing attending surgeons was my sole motivator for mastering these clunky tools.

---

[9] Stanford Medicine. *The Physician Burnout & EHR Burden Report*. Published 2022. Accessed January 27, 2025. https://www.ehrinpractice.com/ehr-failure-statistics.html.

During my medical school years, digital health was even more rudimentary. Everything was paper-based, and lab results were often buried in piles of faxes. At smaller, rural hospitals, when I was on rotations, I spent much of my day running up and down stairs retrieving physical X-ray films for my team to review. Digital resources were scarce, and Internet access was often limited to a bare bones intranet, making research a laborious process. The concept of integrating digital tools into medical education simply didn't exist or wasn't yet a priority.

As of 2025, interoperability remains one of the most pressing challenges in healthcare. Despite years of discussions around standards like FHIR (Fast Healthcare Interoperability Resources), DICOM (Digital Imaging and Communications in Medicine), and HL7 (Health Level Seven International), most systems can still only exchange a fraction of a patient's full record. Critical data such as imaging, labs, and clinical notes often remain fragmented across different platforms, limiting the potential of health innovations, including acceleration of impactful AI.

Next Shift Quick Win

Physicians lose nearly two additional desk-work hours for every hour of face time.[10,11]

Rapid relief plan:

   Day 1: Pull yesterday's audit log and flag the five screens with the longest average dwell time.

   Day 2: Convene one clinician "pain huddle" at shift change; ask for one fix per screen, no blame allowed.

   Day 3: Push the simplest fix live (for example, defaulting to the most common order set) and announce it on the intranet home page; repeat weekly.

The consequences of poor interoperability are far-reaching. Siloed systems force clinicians to rely on incomplete or redundant information, leading to inefficiencies, misdiagnoses, and unnecessary tests. Instead of seamless data sharing, healthcare organizations are often stuck using patchwork solutions: a chaotic mix of legacy

---

[10] Sinsky C, Colligan L, Li L, et al. Allocation of physician time in ambulatory practice: a time and motion study in 4 specialties. *Ann Intern Med*. 2016;165(11): 753–760. doi:10.7326/M16-0961.

[11] Friedberg MW, Chen PG, Van Busum KR, et al. Factors affecting physician professional satisfaction and their implications for patient care, health systems, and health policy. *Rand Health Q*. 2014;3(4).1.

systems, homegrown databases, and commercial platforms with limited standardization.

In an ideal world, a patient's medical history would follow them effortlessly between providers, hospitals, and even across national borders. Instead, data exchange remains highly inconsistent, often requiring manual workarounds like faxing records or forcing patients to carry their medical history themselves. The fax machine is like a zombie in healthcare—it will never die!

The fax machine is like a zombie in healthcare—it will never die!

While FHIR APIs have improved data exchange in recent years, widespread adoption has been slow, and true interoperability is still more of a regulatory requirement than a practical reality. Many health IT vendors (you know who I'm talking about!) and hospital systems continue to operate in walled gardens, hesitant to share data freely due to business incentives, security concerns, or the complexity of standardizing legacy systems. The ultimate goal is not just connecting systems; it is ensuring that the right data reaches the right provider at the right time, in a way that enhances decision-making and improves patient outcomes. Until then, healthcare will continue to struggle with the paradox of the digital age: having more data than *ever* but still lacking the right information when it matters most (see Diagram 2.4).

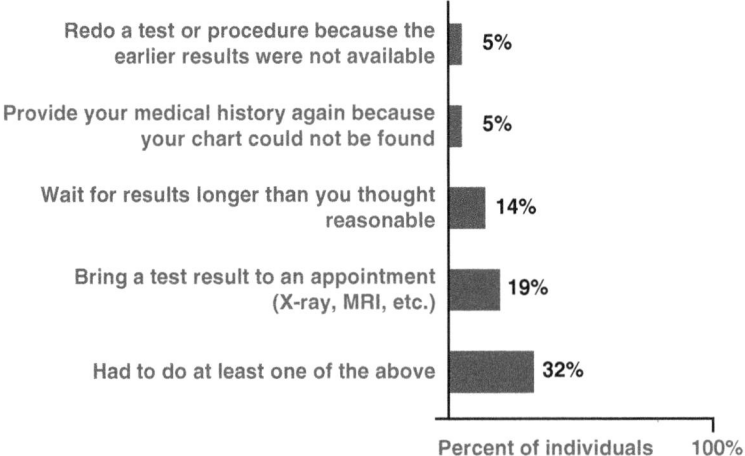

Diagram 2-4: Information Exchange Gap Experienced by Patients in the United States.[12]

---

[12] Office of the National Coordinator for Health Information Technology. *Gaps in Individuals' Information Exchange. Health IT Quick-Stat #56.* HealthIT.gov.

Data Accuracy and Migration Challenges

Transitioning from paper or legacy digital systems to a modern EHR requires careful data migration. Even today, a migration to new or upgraded technologies is no small feat. Early systems lacked the sophistication to handle bulk data migration efficiently, resulting in errors and incomplete records. Ensuring data accuracy during implementation is a time-consuming and costly process. Incomplete or incorrect data transfers can result in medical errors, duplicate records, and lost patient history. Standardizing data formats and conducting thorough validation checks before going live can mitigate risks. Additionally, data integrity must be maintained through ongoing audits and staff training.

While the United States grappled with EHR adoption, other nations took varied approaches. Denmark, with their e-Journalean system, had collected health data on 85 percent of the Danish population by 2021.[13] Although the Internet is present in all African countries, access has been concentrated in urban centers, with no access in most rural centers, where over 80 percent of the population reside.[14] These disparities highlight how access to digital health is shaped by infrastructure, policy, and investment, reinforcing the need for globally scalable solutions.

Inadequate Training and Support

Healthcare providers frequently cite insufficient training as a major pain point during EHR rollouts. You'll remember the 5–30 minutes of casual "training" I received in most places. Organizations often underestimate the learning curve required for staff to become proficient with new systems. Those health systems requiring less than four hours of education for new systems consistently create poor experiences for

---

Published 2019. Accessed January 29, 2025. https://www.healthit.gov/data/quickstats/gaps-individuals-information-exchange.
[13] *eHealth in Denmark—A Stronghold of Coherent and Trustworthy Digital Health Solutions*. The Ministry of Health, Denmark. Published 2021. Accessed January 30, 2025. https://www.ism.dk/Media/637643563459491419/eHealth%20in%20Denmark.pdf.
[14] Fraser HS, Blaya J. Implementing medical information systems in developing countries, what works and what doesn't. *AMIA Annu Symp Proc.* 2010;2010: 232–236. https://pmc.ncbi.nlm.nih.gov/articles/PMC4167769/.

clinicians.[15] Significant jumps in overall user satisfaction occur with every additional hour of initial EHR education received. Additionally, inadequate ongoing support leaves users struggling to troubleshoot issues independently after their training ends.

This lack of structured training creates three key challenges:

1. **Low Engagement:** Clinicians are often expected to figure out digital tools on their own, leading many to delegate tasks to nurses or trainees instead of fully engaging with the system themselves.

2. **Role-Specific Confusion:** Effective implementation requires engaging staff at every level. Providing opportunities for input and fostering a sense of control can make or break a rollout. Robust training programs tailored to specific roles (clinicians, nurses, administrative staff, and IT teams) are essential. A "train-the-trainer" model, where super-users provide peer support, can improve adoption as well.

3. **Lack of Ongoing Support:** Even with strong initial training, clinicians need continuous post-launch support to navigate updates and workflow optimizations. A dedicated IT help desk, along with structured feedback loops for system improvements, ensures long-term success.

Without a strong commitment to education and support, even the best-intentioned digital health initiatives risk falling flat; creating frustration rather than efficiency.

Privacy and Security Concerns

The digitization of sensitive patient data introduces new vulnerabilities. Early EHR systems lacked robust security measures, making them susceptible to breaches and unauthorized access; concerns that created hesitation among providers and patients. Globally, privacy and security are fundamental to everything from national ID systems to banking and healthcare, yet each country has approached these challenges differently, shaping diverse regulatory landscapes (see Diagram 2.5).

---

[15] Miliard M. EHR training is biggest predictor of user satisfaction, experts say. *Healthcare IT News.* Published May 17, 2019. Accessed January 29, 2025. https://www.healthcareitnews.com/news/ehr-training-biggest-predictor-user-satisfaction-experts-say.

Data Protection and Privacy Legislation Worldwide

Legislation
Draft Legislation
No Legislation
No Data

Diagram 2-5: Global Distribution of Privacy and Security Legislation.[16]

No system is impenetrable, and patient information remains one of the most sensitive forms of data. Cybersecurity threats such as ransomware, phishing attacks, and insider breaches pose significant risks to digital health systems. A multi-layered security strategy—including encryption, multifactor authentication (MFA), strict access controls, and regular penetration testing—is essential. Compliance with regulations like HIPAA, GDPR, and other country-specific frameworks is mandatory, but true data security requires a proactive approach, integrating ongoing staff education and continuous threat monitoring.

While these challenges created significant friction in the early days of EHR adoption, they also served as catalysts for innovation. Today's systems are more intuitive, integrated, and secure, reflecting the industry's ability to learn from past mistakes and drive progress. The growing adoption of EHRs marks a fundamental shift: most providers trained today will never practice on paper. They will never have the "privilege" of handwriting admission orders for a complex patient, only to spill their coffee on them! Ideally, they will navigate a digital world where security and usability are paramount, ensuring that technology supports, rather than hinders, patient care.

---

[16] United Nations Conference on Trade and Development (UNCTAD). *Data Protection and Privacy Legislation Worldwide.* UNCTAD ©2025 United Nations. Reprinted with the permission of the United Nations. https://unctad.org/ page/data-protection-and-privacy-legislation-worldwide.

Telehealth: The Early Days

Stories from the Field: Physical Barrier Spotlight[17]

In the heart of rural Virginia, Dr. Wendy Woolley, an emergency department physician, finds herself juggling the impossible. She works in a critical access hospital in Rockbridge County, where the emergency department is a lifeline for its diverse and often underserved population. On any given night, Dr. Woolley might be treating a college athlete from a nearby university, a trucker passing through on Route 81, or a farmer who hasn't seen a doctor in years. The common thread among her patients? They are all navigating a healthcare system riddled with barriers.

"I see patients who see me because they can't drive two hours and can't afford the gas to reach a larger hospital," Woolley says. (**Physical Barrier**) "And others don't come until they're falling apart because they just don't trust doctors."

Her hospital, a 25-bed facility, relies heavily on creative solutions like telemedicine (see Diagram 2.6). But even with these tools, the work is grueling. Staffing shortages mean that Dr. Woolley, along with her team of four nurses and one tech, often stretch themselves thin to meet the needs of their patients. "We're doing everything we can with what we have," she says. "But some days, it feels like we're just putting out fires."

Percentage of visits with telemedicine use, by physician specialty: United States, 2021

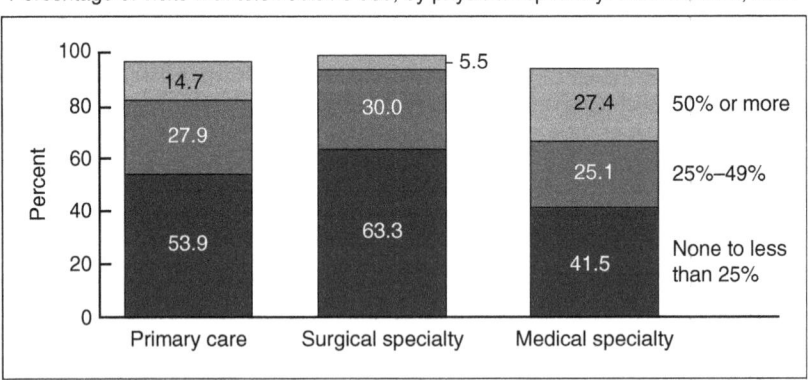

Diagram 2-6: Percentage of U.S. Patient Visits Utilizing Telemedicine by Specialty, 2021.[18]

---

[17] Woolley W. Interview with Sarah Matt. January 22, 2025.
[18] Myrick KL, Mahar M, DeFrances CJ. *Telemedicine Use Among Physicians by Physician Specialty: United States, 2021.* NCHS Data Brief, no. 493. Hyattsville, MD: National Center for Health Statistics. 2024. doi:10.15620/cdc:141934.

In Tazewell County, where she also covers shifts, the challenges deepen. Located on the West Virginia border, the area is steeped in stigma around seeking medical help **(Cultural Barrier)**. Medicaid patients often struggle with cross-state care restrictions, unable to access specialists in neighboring towns. "It's not just about money," she explains. "People don't know how to navigate the system. They don't have a primary doctor they've known their whole lives. And sometimes, they just don't believe in going to the doctor until it's a crisis."

Despite these barriers, Dr. Woolley remains hopeful about the potential of technology like remote robotic surgery and expanded telehealth. Yet she is quick to point out its limitations. "You can't deliver a baby over telehealth," she laughs. "And yes, I've delivered babies in the middle of the night because the nearest labor and delivery ward is 40 minutes away."

Dr. Woolley's story is one of resilience and realism. She's not waiting for the system to fix itself; she's doing what she can, one patient at a time. "Our

You can't deliver a baby over telehealth!

system is broken, no doubt about it," she admits. "But I wake up every day and try to help the people right in front of me. That's all I can do."

For rural hospitals like Dr. Wendy Woolley's, digital health and telehealth are no longer futuristic concepts; they are lifelines. But how did we get here? Understanding telehealth's early days helps us see how far we've come. The origins of telehealth stretch back decades, long before the widespread adoption of smartphones and high-speed Internet. Initially developed as a tool to bridge healthcare gaps in remote and underserved areas, early telemedicine efforts were limited by the available technology and infrastructure. Despite these challenges, these early innovations laid the foundation for the sophisticated virtual care systems we see today.

Telehealth's roots can be traced to the mid-20th century, when healthcare providers began experimenting with communication technologies to deliver care. One of the first documented uses of telemedicine occurred in the 1960s, when NASA collaborated with medical professionals to monitor astronauts' health during space missions. Around the same time, the Nebraska Psychiatric Institute conducted one of the first successful trials of interactive video consultations, connecting psychiatric patients in rural areas to specialists at a central facility.[19] Early initiatives, such as the Royal Flying

---

[19] American Psychiatric Association. *History of Telepsychiatry*. Psychiatry.org. Accessed January 28, 2025. https://www.psychiatry.org/

Doctor Service (RFDS) in Australia, established in 1955, utilized radio communication to provide medical consultations to remote areas. Similarly, Apollo Hospitals in India pioneered telemedicine services to bridge urban-rural healthcare disparities, introducing the first V-sat Tele-Emergency services in Himalayan regions and establishing over 350,000 touchpoints nationwide.[20] Similarly, as of 2020, Rwanda's community-based health insurance scheme covered 88 percent of the population. This scheme meant that Rwanda is one of only a few countries that covered both virtual and in-person care, ensuring that telehealth services would be accessible to a broad segment of the population.[21]

These early experiments demonstrated telehealth's potential to overcome **Physical Barriers** and improve access to care. However, they also highlighted the limitations of the era's technology. The largest obstacles at the time were consolidated into three main areas:

1. **Bandwidth Issues:** Early networks struggled to support the audio and video quality needed for real-time consultations. This often resulted in delays or dropped connections.

2. **Lack of Interoperability** (*notice how this keeps popping up!*): Telehealth systems were siloed, with little to no ability to exchange information across platforms. This created barriers to continuity of care and limited the integration of telemedicine into existing healthcare workflows.

3. **Limited Use Cases:** Early telehealth primarily focused on remote consultations and monitoring, lacking the robust diagnostic and treatment capabilities of modern systems.

Despite its limitations, early telehealth paved the way for today's virtual care innovations. The lessons learned during these formative years informed the development of modern telemedicine platforms, driving advancements in bandwidth, interoperability, and user-centric design. Today, telehealth is a critical component of global healthcare delivery, with increasing adoption across both high-income and low-income nations.

---

psychiatrists/practice/telepsychiatry/toolkit/history-of-telepsychiatry.

[20] Apollo Hospitals. *Apollo Telehealth: Revolutionizing Healthcare Access Through Innovation*. Apollo Hospitals. Accessed January 28, 2025. https://www.apollohospitals.com/apollo-in-the-news/apollo-telehealth-revolutionizing-healthcare-access-through-innovation/.

[21] How digital solutions are improving health equity worldwide. Published June 2022. Accessed January 30, 2025. https://www.weforum.org/stories/2022/06/health-care-equity-digital.

Stories from the Field: Digital Pillar Spotlight[22]

When Dr. Wendy Woolley began working at Rockbridge County's critical access hospital, she quickly saw how limited specialist access put patients at risk. Located in a rural area with only 25 inpatient beds, the hospital lacked neurologists on staff, which was a major problem when patients arrived with strokes, seizures, or other neurological emergencies.

> Before we had Tele-stroke, if a patient came in with stroke symptoms, we were stuck with a tough choice: either transfer them over an hour away or manage them with the limited resources we had. Many of them didn't qualify for the big interventions like clot retrieval, but they still needed care.

Then, the hospital launched a Tele-Neurology program. Suddenly, when Dr. Woolley suspected a stroke or seizure, she could immediately connect with a neurologist via a high-definition video system. The specialist could assess the patient in real-time, review imaging, and help guide treatment, all without the patient leaving the hospital.

> It changed everything. Now, a neurologist comes on the screen. They can move the camera remotely, zoom in to see facial droop or weakness, and guide our team through the exam. If the patient needs urgent transfer, we know immediately. If they can stay, the tele-neuro team rounds on them daily, right from the screen. That means fewer unnecessary transfers and better care right here.

For rural hospitals like Wendy's, telehealth isn't just a convenience, it's a lifeline. Without it, patients would continue to fall through the cracks, waiting too long for specialist care, or worse, never getting it at all.

The adoption of telehealth has varied globally, influenced by factors such as infrastructure, policy, and healthcare needs. In India, Apollo TeleHealth has transformed healthcare access through innovation, impacting the lives of more than 20 million individuals by providing virtual consultations and remote healthcare services.[23] In Australia, the RFDS performed 55,930 telehealth consultations in the 2023–2024 period, highlighting the critical role of telemedicine in providing care

---

[22] Woolley W. Interview with Sarah Matt. January 22, 2025.

[23] Apollo TeleHealth. *Telemedicine Services*. Apollo Hospitals. Published 2000. Accessed January 22, 2025. https://www.apollohospitals.com.

to remote populations.[24] These examples demonstrate the potential of telehealth to overcome geographical barriers and improve healthcare accessibility.

The COVID-19 pandemic served as a catalyst for telehealth adoption worldwide. Policy changes, such as expanded reimbursement models, allowed telehealth to flourish. What had been a fringe offering turned into a primary mode of care for millions. Patients relied on virtual visits for everything from routine checkups to critical consultations. It was during this time that digital health shed its label as a "nice-to-have" and emerged as a necessity. The potential I had once doubted was now undeniably clear: digital transformation was not just possible; it was now indispensable.

Telehealth Gains How-To

Fifty-four percent of Americans have used telehealth; almost 9 in 10 say it made care easier.[25]
Make the surge stick:

1. **Reroute Low-Acuity Follow-ups:** Auto-offer a video slot before an in-person slot for any visit with a normal vital-sign set.

2. **Embed a 30-Second "Tech Check" at the start of every call:** Camera, sound, privacy reminder, ready.

3. **Track "Video-First" Completion:** Post the weekly no-show delta between virtual and in-person visits on the ops dashboard; share wins with schedulers every Monday.

A Quiet Revolution in Personal Wellness

While telehealth and EHRs were created to improve efficiency for providers; patients who are now more of a consumer of healthcare needed their own autonomy. The rise of consumer health devices and apps has fundamentally changed how individuals monitor and engage with their well-being. What began with simple pedometers and basic heart rate monitors has evolved into an expansive

---

[24] Royal Flying Doctor Service. *90 Years of Unparalleled Service*. Royal Flying Doctor Service. Published 2017. Accessed January 22, 2025. https://www.flyingdoctor.org.au/news/90-years-unparalleled-service.

[25] Positively Impacting Society. *2024 National Telehealth Survey Results*. Published January 2024. Accessed January 27, 2025. https://pos.org/2024-national-telehealth-survey.

ecosystem of tools designed to empower users with actionable health insights. While early devices faced skepticism and technological limitations, they laid the groundwork for the current explosion of wearable health technologies and apps. I'm sure as you read this, you are personally hooked up to at least one consumer health device! For me, I'm sitting here with my Oura ring on, and my cell phone within arm's reach, while typing on my computer (see Diagram 2.7).

The first wave of wearable devices, such as pedometers, offered a straightforward proposition: track your steps and encourage physical activity. Though basic in functionality, these devices succeeded in promoting awareness around movement and personal fitness. Their accessibility made them popular among consumers looking to make small but meaningful lifestyle changes. Similarly, early heart rate monitors, often integrated into chest straps or standalone devices, found favor with athletes and fitness enthusiasts seeking to optimize their training. Despite their simplicity, these tools marked the beginning of consumer shift toward real-time health monitoring. You may remember that my tiny Fitbit had a huge impact on my vision of the future of connected care.

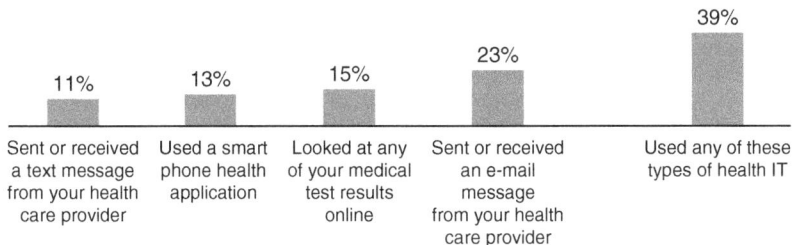

Proportion of individuals who report using selected types of health IT in 2013.

11%	13%	15%	23%	39%
Sent or received a text message from your health care provider	Used a smart phone health application	Looked at any of your medical test results online	Sent or received an e-mail message from your health care provider	Used any of these types of health IT

Diagram 2-7: Proportion of Individuals Who Report Using Selected Forms of Health IT in 2013.[26]

With the advent of smartphones, the next step in the evolution of consumer health came in the form of health apps. These apps, ranging from calorie trackers to guided meditation programs, initially met with mixed reception. Critics dismissed them as gimmicky, while others

[26] Patel V, Barker W, Siminerio E. Disparities in Individuals' Access and Use of Health IT in 2013. *ONC Data Brief*, no. 26. Washington, DC: Office of the National Coordinator for Health Information Technology; June 2015. Accessed January 28, 2025. https://www.healthit.gov/sites/default/files/briefs/oncdatabrief26june2015consumerhealthit.pdf.

lauded their potential to democratize access to health resources. Over time, certain apps began to stand out by offering tangible benefits, such as facilitating weight loss, improving sleep habits, or enhancing mental well-being. Apps like MyFitnessPal gained widespread adoption by combining user-friendly interfaces with data-driven insights, allowing individuals to track their food intake and exercise routines with unprecedented ease.

Perhaps one of the most significant milestones in this evolution has been the integration of wearables and apps. Devices like Fitbit and Apple watch combined step tracking, heart rate monitoring, and app connectivity to provide users with a more comprehensive view of their health. These tools offered not only raw data but also actionable insights, such as reminders to move, tailored fitness goals, and sleep quality reports.

Although early iterations had limitations, such as inaccurate step counts or difficulties with syncing data, they still managed to engage users in a way that fostered healthier habits. Despite their growing utility, these early health devices and apps were not without their challenges. Questions about data accuracy, privacy, and the actual long-term impact on health behaviors persisted. Yet this did not overshadow their value as catalysts for a new era of health consciousness. They paved the way for more advanced iterations, inspiring innovation and signaling a future where technology and health are seamlessly intertwined.

The consumer health revolution continues to gain momentum, fueled by advancements in wearable technology, artificial intelligence, and data integration. What began with simple step counters has evolved into a sophisticated network of tools that empower individuals to take charge of their health. We are now on a journey that is transforming how we think about wellness, prevention, and the role of technology in personal health.

Stories from the Field: Digital Barrier Spotlight[27]

For many patients, digital health wasn't just about convenience, it was the difference between getting care and going without. Dr. Wendy Woolley relayed to me her encounter with a diabetic patient we'll call John.

John had lived his whole life in a rural community, working long hours as a mechanic. He had diabetes, but because of the long

---

[27] Woolley W. Interview with Sarah Matt. January 22, 2025.

distance to the nearest specialist and his reluctance to take time off work, he hadn't seen a doctor in years. When he finally showed up at Wendy's emergency room, he was in full-blown diabetic ketoacidosis, a life-threatening condition.

"The thing is," Wendy said, shaking her head, "If he had access to even a simple remote monitoring device that could alert him to his dangerously high blood sugar levels, we could have caught this earlier. But for patients like John, the healthcare system only shows up when they're in crisis."

John survived, but his case reinforced what I had started to realize: technology wasn't just about efficiency. It had the power to prevent emergencies, reduce suffering, and change lives. This was no longer about futuristic healthcare; this was about bringing essential care and **Digital Access** to those who need it most, today.

Next Shift Quick Win: Wearables That Work by Monday

Why It Matters: Regular Fitbit engagement cuts overweight risk by 46 percent,[28] a single inexpensive tracker can turn passive patients into data-sharing partners.

Three Things the Night Team Can Do Right Now:

1. **Loan a Tracker at Discharge:** Preload a QR code that opens a one-page setup guide in the portal.

2. **Set an Automatic Nurse Call:** Do this for any patient whose seven-day step count falls 30 percent below baseline. Use it as a "digital nudge" rather than a reprimand.

3. **Publish an Anonymized Steps Scoreboard on the Breakroom Monitor Every Friday:** Celebrate the top three movers with a coffee gift card.

Meaningful Use and U.S. Policies Driving EHR Adoption

With EHRs struggling to gain widespread adoption due to cost, provider resistance, and lack of interoperability, policymakers stepped in. The U.S. government saw EHRs as essential to the future of healthcare, leading to the introduction of Meaningful

---

[28] Fitbit Usage Statistics. *Market.us Media.* Accessed January 28, 2025. https://media.market.us/fitbit-usage-statistics.

Use and MACRA (Medicare Access and CHIP Reauthorization Act). These programs were designed to push hospitals and clinics toward digital adoption through financial incentives and penalties (see Diagram 2.8). The introduction of the HITECH (Health Information Technology for Economic and Clinical Health) Act in 2009 was a pivotal moment in the digitization of U.S. healthcare. At its core, the act introduced the concept of *meaningful use* (MU), incentivizing healthcare organizations to adopt EHRs and use them in ways that improved patient outcomes. It established the MU program, a phased initiative designed to promote:

- **Improved Quality, Safety, and Efficiency:** By replacing paper records with digital systems, providers gained tools to reduce medical errors and improve coordination.

- **Reduction of Health Disparities:** MU aimed to make healthcare more equitable, addressing gaps in care for underserved populations.

- **Patient and Family Engagement:** Empowering patients with access to their health data became a central tenet, encouraging active participation in care.

- **Privacy and Security:** Safeguarding patient information was essential as healthcare data transitioned to digital platforms.

MU incentivized compliance by offering financial bonuses to providers who met specific criteria, such as electronic prescribing and data sharing. Noncompliance, on the other hand, eventually resulted in penalties, ensuring widespread participation. While MU laid the groundwork for EHR adoption, subsequent policies like MACRA and MIPS (Merit-based Incentive Payment System) expanded its scope and impact. Together, these policies transformed the healthcare reimbursement landscape, linking payments to quality and value.

- **MACRA:** Passed in 2015, MACRA introduced value-based reimbursement models to replace the traditional fee-for-service approach. It emphasized outcomes over volume.

- **MIPS:** As part of MACRA, MIPS consolidated MU with other programs like the Physician Quality Reporting System (PQRS) into a single framework. Providers were measured on four pillars: quality, cost, improvement activities, and promoting interoperability.

Year	Hospitals	Office-based Physicians
2008	9%	17%
2009	12%	22%
2010	16%	28%
2011	28%	34%
2012	44%	40%
2013	59%	48%
2014	97%	74%
2015	96%	78%
2016	96%	78%
2017	96%	80%
2018	96%	79%
2019	96%	72%
2021	96%	78%

Diagram 2-8: Hospital and Physician Adoption of Electronic Health Records in the United States from 2008 to 2021.[29]

These programs collectively reinforced the need for EHRs to enhance quality, safety, and efficiency, not just serve as digital record-keeping systems. The policies surrounding meaningful use (MU), MACRA, and MIPS achieved several critical objectives:

1. **Driving Technology Adoption:** By 2021, 96 percent of hospitals and 78 percent of office-based practices in the United States had adopted EHRs.[30]

2. **Shaping Care Delivery:** The emphasis on interoperability and patient engagement improved care coordination and access to health information.

3. **Improving Outcomes:** Metrics tied to these programs have driven providers to focus on managing chronic disease, providing preventative care, and reducing hospital readmissions.

In theory, these regulations sounded promising: they aimed to incentivize hospital systems to modernize and improve patient care. However, their actual implementation faced significant resistance and

---

[29] Patel V, Barker W, Siminerio E. *Disparities in Individuals' Access and Use of Health IT in 2013.* ONC Data Brief, no. 26. Washington, DC: Office of the National Coordinator for Health Information Technology; June 2015.
[30] Office of the National Coordinator for Health Information Technology. *National Trends in Hospital and Physician Adoption of Electronic Health Records.* HealthIT.gov. Published 2021. Accessed January 29, 2025. https://www.healthit.gov/data/quickstats/national-trends-hospital-and-physician-adoption-electronic-health-records.

unintended consequences from healthcare vendors and healthcare systems:

- **Burden on Providers:** Many clinicians reported that MU and MIPS added substantial administrative workload, diverting time away from direct patient care.

- **Interoperability Gaps:** While EHR adoption surged, the absence of universal data-sharing standards hindered the seamless exchange of patient information across different systems.

- **Balance in Adoption:** Smaller practices and rural providers often struggled to meet the requirements due to financial and technical constraints.

- **Usability Challenges:** EHR vendors were required to quickly certify or recertify their systems so hospitals/providers could qualify for incentives. The aggressive development timelines often forced shortcuts, typically at the expense of usability enhancements and thorough quality testing.

Policy influencers like the American Medical Association (AMA) and the Electronic Health Record Association (EHRA) worked to refine and clarify the legislation. They provided commentary to regulators to better serve patients and ensure the regulations were realistically achievable within the set timelines. However, not all recommendations were incorporated. These challenges highlight the critical need to align policy with real-world clinical workflows and technology, a lesson that continues to shape healthcare innovation today.

Stories from the Field: Digital Barrier Spotlight

During MIPS/MACRA, and the three phases of the meaningful use program (MU1, MU2, and MU3), I was leading product management within a technology organization focused on creating EHRs and their extensive surround. Each time regulations changed, it was very challenging. It felt like as soon as we finally "finished" our work on MU2, we were already starting the work on MU3. Then when we had finally finished that, suddenly, we were incorporating new requirements for MIPS and MACRA. The pace was ruthless.

My last memorable push was when I was driving our patient engagement strategy. Our main engagement product at the time was our patient portal. We had a grand vision of making our portal consumer grade, a beautiful experience that patients would be delighted to use. My team had researched all the latest wearables, as well as consumer health and wellness trends. We had done extensive comparisons of feature sets across patient engagement solutions around the world. We had created surveys and spoke to patients of all ages and stages about their pain points and experiences. It was a great time to be in product management; we were solving problems and focusing on our users.

The previous MU push had left some of the portfolio with areas that still needed software updates at the very foundation and there were a lot of must-haves that needed to have bugs ironed out. Our development team had been deployed to eliminate these issues. This gave us time to plan and pour over screen design and get amazing feedback. We even got the go ahead to use an outside firm to build out the first workflows to accelerate our path.

Then a new set of regulations was released. We knew they were coming, but sometimes when policy is passed down, you never really know when they will release the information, how long a comment period will be, or when to expect a final rule. This level of uncertainty meant that the final outcome could be very different than what the original documentation implied. This time the changes were big, and not just for the portal, but the whole portfolio. I had to give my team a serious pep talk. "We've done amazing work, and we've made some amazing designs. We're going to put these on hold for the moment and dig into the required regulations. We aren't sure exactly what the final rule will be, but given the short delivery window, we need to start now. Don't worry, we'll get back to this consumer grade experience in no time."

Like many other solutions in the market, our consumer grade workflows, our relentless focus on the user, and our vision of something better didn't happen. Instead, we were barely able to add the necessary regulatory features in time. Most of the time we didn't have time to speak with users or focus on design, since the policy was so specific and time-consuming to implement. There was nothing delightful about the experience we created, and it tore at us.

Today I don't know anyone who loves their patient portal. They have certainly improved incrementally, but it's been over a decade since I had to put those visions of delight on hold. Up to this point across the portal space not much has truly changed since that time.

> **Fact Check**
>
> 1. MACRA's two-track payment system encouraged physicians to migrate to alternative payment models.[31]
> 2. [MIPS] compliance cost $12,800 per physician per year, and physicians spend 53 hours per year on MIPS-related tasks. This is the equivalent of a full week of patient visits.
> 3. Physicians caring for more patients from historically marginalized racial and ethnic groups were more likely to receive low scores [for MIPS], despite providing high-quality care.[32]

Global Regulatory Innovations

While the United States relied on financial incentives and penalties to drive EHR adoption, other countries took a different path. Some focused on privacy and security (General Data Protection Regulation—GDPR—in Europe), while others prioritized interoperability (India's NDHB) or telehealth reimbursement (Australia). These international examples show that digital transformation isn't one-size-fits-all. As digital health continues to reshape global healthcare systems, nations around the world are adopting innovative regulatory frameworks to address challenges in privacy, accessibility, and integration. From the stringent privacy standards of Europe to the unified digital health strategies in India, these efforts illustrate how thoughtful policy can advance healthcare delivery and foster the **Trust/Knowledge Pillar** in emerging technologies.

In Europe, the implementation of the GDPR in 2018 marked a turning point in the protection of personal health data. Designed to enhance transparency and accountability, GDPR requires explicit patient consent for data use and gives individuals the right to access, amend, or delete their information. For healthcare, this has fostered greater trust in digital platforms, ensuring that patients retain control over their sensitive health data. However, its rigorous requirements have created compliance challenges, particularly for smaller healthcare providers and technology startups that may lack the resources to

---

[31] RAND Corporation. *MACRA's Effects on Medicare Payment Policy and Spending.* Published 2017.

[32] American Medical Association. *It's Time to Revamp Medicare's Broken MIPS Program.* Published July 2024. Accessed January 29, 2025. https://www.ama-assn.org/practice-management/payment-delivery-models/it-s-time-revamp-medicare-s-broken-mips-program.

navigate its complexities. Despite these hurdles, GDPR has set a global benchmark for data privacy and continues to influence healthcare data policies worldwide.

In Australia, the expansion of telehealth under the Medicare Benefits Scheme (MBS) demonstrates the critical role of reimbursement policies in driving adoption. The Australian government leveraged the MBS to subsidize virtual consultations, enabling patients to access care from general practitioners, specialists, and allied health professionals without **Financial Barriers**. Both real-time and asynchronous telehealth services are covered, making it one of the most comprehensive reimbursement frameworks globally. During the COVID-19 pandemic, these policies proved invaluable, with more than 80 percent of Australians accessing telehealth services in some form. By bridging the gap between rural and urban healthcare delivery, MBS has become a model for how policy can enhance access while adapting to the needs of a modern healthcare system.[33]

India offers another compelling example of innovation with its National Digital Health Blueprint (NDHB). Launched in 2019, the NDHB envisions a unified digital health ecosystem, supported by the creation of a unique health ID for every citizen. This ambitious framework aims to integrate public and private healthcare systems, standardize data exchange, and enhance interoperability. By building on the foundations of the Ayushman Bharat Digital Mission, the NDHB has laid the groundwork for a system that brings together telemedicine platforms, electronic health records, and digital applications under a single, accessible framework.[34] In a country with over 1.4 billion people, this initiative is addressing the longstanding barriers to efficient healthcare delivery by breaking down **Digital Barriers**.

Together, these examples illustrate the transformative potential of regulatory innovation. Europe's emphasis on privacy, Australia's success in telehealth reimbursement, and India's commitment to interoperability highlight the diverse approaches nations are taking to meet the challenges of digital healthcare. These efforts not only improve patient outcomes but also ensure that the systems of tomorrow are built on trust **(Trust/Knowledge Pillar)**, accessibility, and inclusivity.

[33] Department of Health and Aged Care. *Permanent Telehealth to Strengthen Universal Medicare*. Australian Government. Published December 13, 2021. Accessed January 29, 2025. https://www.health.gov.au/ministers/the-hon-greg-hunt-mp/media/permanent-telehealth-to-strengthen-universal-medicare.

[34] Jain N, Singh H. India's Ayushman Bharat Digital Mission (ABDM): Towards a unified digital health ecosystem. *J Family Med Prim Care*. 2023;12(1):153–157. doi:10.4103/jfmpc.jfmpc_1073_22. https://pmc.ncbi.nlm.nih.gov/articles/PMC10064942.

Key Lessons from Global Digital Health Regulations:

- **GDPR (Europe):** Ensured privacy but increased compliance burden.

- **MBS (Australia):** Government reimbursement helped scale telehealth.

- **NDHB (India):** Created a unified digital health system for more than 1B people, but with disparities.

- **MIPS/MACRA (United States):** Pushed EHR adoption but increased provider burnout.

Building Momentum for the Future

Reflecting on the transformative changes brought by digital healthcare, it's astounding to see how far we've come. The problem is no longer the technology—there's an abundance of advanced tools, algorithms, and ever-evolving capabilities. The real challenge lies in ensuring these technologies remain patient-centered and user-focused. Whether the user is a healthcare professional navigating complex workflows or a patient trying to access their medical records, the experience must address their unique needs. Too often, solutions designed for one audience are awkwardly retrofitted for another, leading to inefficiencies and frustrations. Instead of building "tech for tech's sake," we must develop tools that solve problems in ways that were previously impossible.

In the United States, regulatory frameworks like MU, MIPS, and MACRA accelerated the adoption of digital tools, but at a cost. While these policies drove widespread implementation, we've seen how they often outpaced usability and efficiency. I recall the frenzy of meeting the requirements of MU2 and MU3. Features were hastily added to meet compliance deadlines, frequently sacrificing workflow and provider experience. These missteps remind us that policy and technology must evolve in tandem, ensuring that healthcare professionals aren't burdened by poorly designed systems, but rather empowered by them.

Globally, other nations offer valuable insights into what's possible. Denmark has achieved remarkable success in healthcare data transparency, enabling seamless access to unified medical records for patients and providers.[35] Meanwhile, India's NDHB is tackling the immense challenge of building a comprehensive health ecosystem for over a billion citizens. They are proving that scalable innovation can be achievable in complex environments, but disparities must be intentionally overcome.[36] These examples show us that technology can do more than digitize records; it can close care gaps, improve outcomes, and create truly borderless healthcare systems.

From Denmark's seamless data-sharing model to India's ambitious digital health infrastructure, these global innovations are pushing healthcare forward. But for all this progress, a fundamental issue remains—access isn't equal, and not everyone is benefiting from these advancements. So, who truly gains from digital health innovation?

Who truly benefits from digital health innovation?

The invitation is yours: choose one barrier in your clinic, startup, company, or community and pilot a small fix; measure the result, share the lesson, and inspire someone else to iterate. Each micro-success adds another plank to the digital bridge. Together we can replace distant promises with everyday access. Keep turning the pages and keep testing new ideas in the real world.

🔧 Startup Builder's Box: Building for the Digital Barrier

If you're not solving workflow, you're part of the problem.

1. **Fit into the Flow:** If your product adds clicks or disrupts charting, you will lose the room. Work within existing systems.

2. **Train Fast, Train Once:** One Friday lunch-and-learn should be enough. Design for minimal onboarding and fast adoption.

3. **Survive the Standards Soup:** If your tool doesn't speak HL7, FHIR, or another standard, it won't scale. Interoperability is your price of admission.

(continues)

---

[35] *eHealth in Denmark—A Stronghold of Coherent and Trustworthy Digital Health Solutions.* The Ministry of Health, Denmark. Published 2021. Accessed January 30, 2025. https://www.ism.dk/Media/637643563459491419/eHealth%20in%20Denmark.pdf.

[36] Gopalakrishnan S, Sujatha R. India's National Digital Health Blueprint: The road ahead for digital health in India. *J Family Med Prim Care.* 2022;11(8):3980–3985. doi:10.4103/jfmpc.jfmpc_1167_22.

> **Startup Builder's Box: Building for the Digital Barrier (continued)**
>
> 4. **Show ROI Early:** Prove value in weeks, not quarters. Time saved, errors avoided, throughput increased—make it quantifiable.
>
> 5. **Respect the Burnout:** No clinician has time for complexity. If your solution depends on perfect compliance, simplify it again.
>
> Healthcare is complex. Your product shouldn't be.

Leaders' End-of-Chapter Action Checklist: Chapter 2: "The Digital Pillar"

LEADER	HIGH-IMPACT ACTION TO STRENGTHEN DIGITAL ACCESS
❑ Board Director	Approve a roadmap to close digital disparities; dedicate 10% of capital to rural broadband; publish quarterly progress
❑ Chief Executive Officer	Form a Digital Governance Council within 30 days; release a 12-month interoperability scorecard
❑ Chief Information Officer	Show live API counts, response times, and downtime minutes on the intranet; flag weekly outliers
❑ Chief Health Information Officer	Require FHIR format on every outbound message; reach full staff privacy certification in six months
❑ VP Clinical Operations	Launch an AI triage chatbot; retune bias and escalation logic every month
❑ VP Nursing and Patient Education	Push micro-learning on portal use and digital consent; achieve 80% staff completion in six months
❑ VP Data and Analytics	Audit fairness, privacy, and explainability in every triage model; present findings to the Council
❑ Telehealth Program Manager	Open every virtual visit with a 30-second privacy promise; sample 10 sessions each month
❑ Patient Experience Manager	Add a one tap emoji feedback bar; target a 10% rise in Net Digital Experience within a year
❑ Community Health Worker Supervisor	Run monthly digital cafés that teach portal logins; lift activation 15% in six months
❑ Director of Snacks and Morale	Serve cookies iced with QR codes that link to the digital roadmap; celebrate milestones with sugar and metrics

The Missing Link in Digital Health: Systems, Not Just Technology

When people talk about healthcare innovation, the conversation often revolves around technology: artificial intelligence, telemedicine platforms, wearables, and mobile health apps. Yet history, and current evidence, suggests that impactful healthcare advancements rarely succeed on technology alone. They flourish because they are embedded in functional systems that align policy, funding, workflows, and human capital. In other words, if digital health were a rock band, the technology would be the lead singer, but the real magic happens because of the drummer, the bassist, and the person who booked the gig. Technology can certainly enhance care delivery, but it cannot fix a broken system.

Even the most advanced telemedicine platform falters if physicians aren't reimbursed for virtual visits. AI-driven diagnostics can indeed speed and refine diagnoses, but only if interoperable systems are in place so that hospitals can integrate algorithmic insights into everyday care. Wearables that generate real-time health data have enormous potential for remote monitoring, unless that data remains siloed and never meets the patient record. The leap from bright idea to widespread adoption relies on a stable infrastructure supporting regulations, payment, and clinician workflow integration.

This reality explains why many digital health startups fail: not because their technology is subpar, but because they do not or cannot

Technology can certainly enhance care delivery, but it cannot fix a broken system.

integrate into a larger healthcare system. Countries like Kenya, India, and Estonia demonstrate how telehealth and AI can scale to national levels when they are integrated into strong systems with coordinated policy, robust financing, community engagement, and interoperable data standards.

Digital Health Ecosystems

The preceding chapters touched on the five pillars of healthcare access. When we consider a full healthcare ecosystem, the system must address all these factors to ensure citizens/patients can get the care they need (see Diagram 3.1).

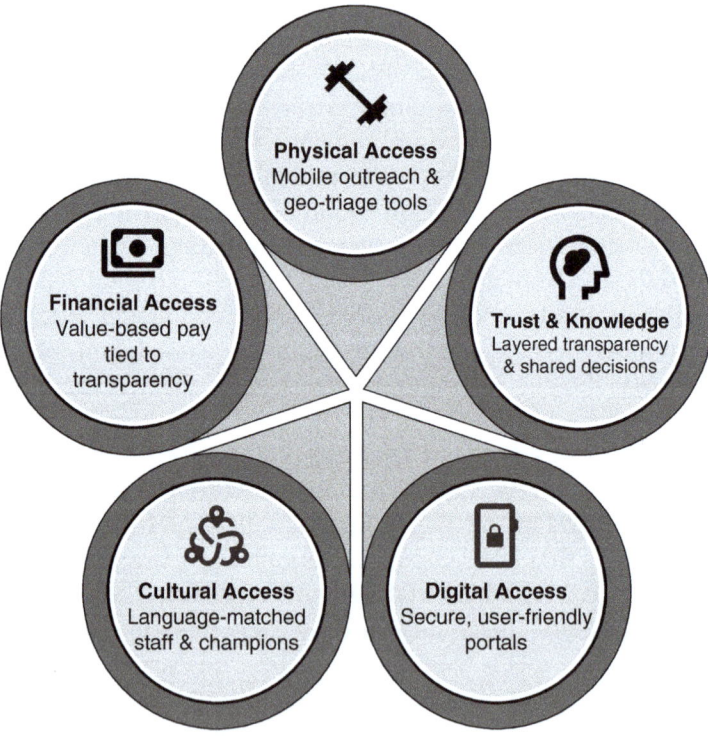

Diagram 3-1: Five Pillars Snapshot.

However, unlike a patient's view of the five pillars, a system needs to consider some slight differences as it works to provide needed care to its constituents. But what is a "healthcare system"? A healthcare system can be any size and any healthcare setup, from an Integrated Delivery Network like we see in the United States, to a county, a state, a full country, or even larger. It all depends on the lens with which you are analyzing the system. When we consider the various parts of the five pillars of access, there are elements that repeatedly appear in the success stories and the cautionary tales of digital health initiatives at the system level. To make these points clearer, the following concise framework recurs throughout this chapter. You won't see the **Physical Pillar** in this list, because it varies so much from region to region. However, pay special attention to how these components relate to the larger pillars of access (see Diagram 3.2):

Financial Pillar:

- **Regulatory Alignment**
  - Flexible policies governing telehealth, AI approval, and cross-border services
  - Clear data privacy/security mandates that build public and provider trust
- **Sustainable Reimbursement**
  - Payment models (public or private) that incentivize digital care delivery
  - Long-term commitment from payers (government or insurers) so startups and hospitals can plan

Digital Pillar:

- **Interoperability and Infrastructure**
  - EHRs, hospital systems, and telehealth platforms capable of exchanging data seamlessly
  - Adequate broadband coverage, reliable electricity, and standardized data protocols
- **Clinical Buy-In and Workforce Training**
  - Providers who understand and trust new tools, with ongoing training and support
  - Clear demonstration of technology's clinical value and usability in real-world settings

Cultural, Trust/Knowledge Pillars:

- **Community Engagement and Digital Literacy**
    - Patients who can access and understand digital health tools (**See Chapter 1 if you need a refresher on digital literacy*)
    - Local champions (e.g., community health workers, government leaders) to promote adoption

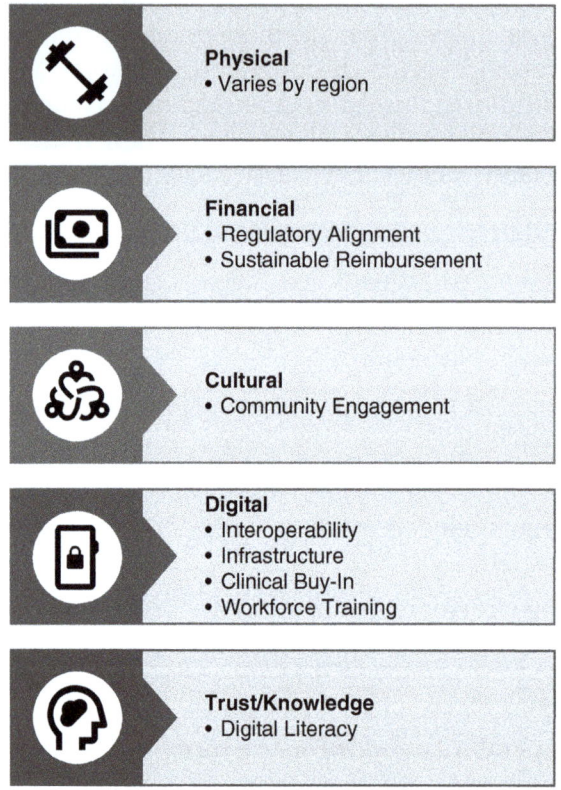

Physical
- Varies by region

Financial
- Regulatory Alignment
- Sustainable Reimbursement

Cultural
- Community Engagement

Digital
- Interoperability
- Infrastructure
- Clinical Buy-In
- Workforce Training

Trust/Knowledge
- Digital Literacy

Diagram 3-2: Five Pillars of Healthcare Access and the Most Influential System-Level Components.

Throughout this chapter, you see how each of these pillars and its components has supported or undermined digital health ventures. By keeping this framework in mind, you can better understand why certain interventions thrive, while others that appear promising on paper collapse under real-world pressures. This chapter also explores why digital health succeeds in some countries and fails in others; not because of better technology, but because of stronger healthcare systems that enable, support, and sustain innovation. The chapter examines the pitfalls that have led to the failure of systems alongside global success stories that prove digital health isn't just about what you build; it's about where and how you build it.

The Hype vs the Reality of Health Tech

Sometimes health tech feels like a terrible 1980s haircare advertisement. It promises to give you hair down to your butt, with glossiness, strength, and the envy of all around you. As that 80s kid with frizzy hair and an outfit picked out by my mother, I can tell you the hype did not actually deliver. The trick with healthcare technology, however, is that careers are made or broken on a decision. Have you ever heard an exec who moved an entire healthcare system to Epic (one of the biggest, and most expensive EHRs in the world) complain loudly? Probably not. They just spent millions and countless human hours on an implementation. If it's a failure, so are they. Amazing how that works. So, what kinds of promises have been made, and where does it differ from real-world adoption? Have the huge promises delivered? See Diagram 3.3.

INNOVATION	HYPE	REALITY
AI	Fully Automated	Requires Human Oversight[a]
Telehealth	Universally Accessible	Reimbursement and Adoption Challenges
Wearables & Remote Monitoring	Reducing Hospitalizations	Issues with Data Integration

[a]To the "tech bro" (I say that with dripping sarcasm, because you were not nice to my colleagues) I met while researching this chapter. Yes, agentic AI "can do all the things," no human needed. But in healthcare, the system will demand a human in the loop until they have solid trust **(Trust/Knowledge Pillar)**. So, once you deliver trust, yes, the need for humans in the loop will continue to decrease and perhaps disappear.

Diagram 3-3: Hype vs Reality of Healthcare Innovations.

Uganda: When Too Many Pilots Sink the System

Governments, too, are susceptible to overpromising. Nowhere was this more apparent than in Uganda, where multiple digital health initiatives operated independently, creating a confusing, inefficient environment for hospitals and clinics (see Diagram 3.4). By 2020,

Uganda had over 50 active health innovation pilots, a phenomenon many dubbed "Pilotitis."

Next Shift Quick Wins: Stop Saying "Pilot" and Start Saying "Phase 1"

Why It Matters: "Pilot" implies throwaway. "Phase 1" implies continuity.

Next shift action:

■ Label your rollout docs as "Phase 1."

■ Include the budget and success metrics for Phases 2 and 3.

■ Tell the staff: "We're starting small, but planning big."

For years, donor-funded digital projects poured in, each promising to improve healthcare delivery in its own way. Yet instead of a unified system, hospitals and clinics were forced to juggle fragmented, overlapping platforms that did not communicate with each other. Doctors struggled with tools that lacked interoperability, a glaring **Digital Barrier**. Meanwhile, every new pilot imposed additional training requirements on already overextended clinicians, undermining clinical buy-in and further creating a **Digital Barrier**. Worse still, many pilots disappeared once external funding dried up, highlighting an absence of sustainable reimbursement and consistent local financing plans. This built up the **Financial Barrier**. Basically, it was like a buffet where every chef brought their own dish, but nobody checked if the flavors worked together. The result? A whole lot of effort and a very confusing meal.

Pilotitis!

In response, Uganda's Ministry of Health took an unprecedented step: they paused all new digital health pilot programs. This regulatory decision was not an indictment of innovation but an acknowledgment that uncoordinated, one-off projects often do more harm than good. The government set out to standardize efforts and prioritize interoperability, ensuring that successful solutions were fully integrated before new ones were introduced.

Uganda's experience reminds us that even the most advanced technology cannot fix a broken system unless it is thoughtfully implemented, harmonized, and designed for long-term sustainability. Equally vital is community engagement and digital literacy: donor-driven pilots often lacked the grassroots support and local training needed to endure. By temporarily halting new pilots, Uganda forced itself and its donor partners to address the underlying system gaps that had led to fragmented information "islands." This case illustrates the critical lesson that robust digital health ecosystems must rest on all five pillars, rather than on technology alone.

Fact Check:[1] Focus on the Population of Uganda

1. 15.8 percent of the population own a digitally enabled phone (18.1% of females, 13.4% of males).

2. 0.9 percent of the population have access to the Internet.

3. 2.3 percent of Ugandans who own mobile phone use routers for Internet access.

PILLAR	UGANDA'S SITUATION	IMPACT
Financial Pillar Regulatory Alignment 	Initially lacked a unifying regulatory approach. Each donor-funded pilot followed its own path, creating a fragmented landscape. The Ministry of Health later intervened, imposing a moratorium on new pilots.	Uncoordinated environment led to overlapping programs; the government's subsequent moratorium forced standardization efforts, illustrating the key role of national policy.
Sustainable Reimbursement 	Most pilots relied on short-term external donor funding; no stable local financing or reimbursement structures were in place. Once funding dried up, many projects vanished.	Lack of a long-term payment models meant promising digital tools disappeared, leaving hospitals with fragmented systems and an underutilized workforce trained in soon-abandoned platforms.

(continues)

---

[1] Ministry of Health, Uganda. *Health Information & Digital Health Strategic Plan 2020/21–2024/25*. Uganda Ministry of Health website. Published March 1, 2023. Accessed February 15, 2025. https://library.health.go.ug/sites/default/files/resources/Health%20Information%20%26%20Digital%20Health%20Strategic%20Plan_01032023_Print_Corrected.pdf.1.

(Continued)

PILLAR	UGANDA'S SITUATION	IMPACT
Digital Pillar Interoperability & Infrastructure	Different pilot projects brought in diverse platforms that rarely communicated, creating numerous "information islands." Internet and power infrastructure varied across regions; many tools were incompatible with existing EHRs or national data standards.	Clinical chaos and duplication of effort for doctors juggling multiple logins, data formats, and training modules. Overall, the infrastructure did not support a single, cohesive digital health ecosystem.
Clinical Buy-In & Workforce Training	Providers were introduced to repeated waves of new platforms, each requiring separate training sessions and workflows. Healthcare workers grew frustrated with constant turnover of solutions and minimal continuity.	Burnout and skepticism among clinicians undermined adoption. Without stable systems or integrated workflows, many solutions failed to gain lasting traction, hindering overall digital health progress.
Cultural Pillar Community Engagement **Trust/Knowledge Pillar** Digital Literacy	Many donor initiatives focused on rapid technology deployment with limited grassroots involvement. Few programs invested in ongoing education for local staff or patients to build long-term digital literacy.	Low local ownership contributed to program collapse once external partners left. Communities saw little incentive to trust or maintain technologies that might vanish without warning.

Diagram 3-4: Uganda's System Pillars.

Japan's Digital Health Transformation

Japan's pursuit of a "Medical DX" future showcases the interplay of multiple factors required for effective digital health adoption at the system level. Although the country boasts a longstanding reputation for technological advancement, it has found that the actual scaling of remote care and AI-based solutions are more complex than anticipated. By examining Japan's situation through the pillars, we gain a clearer picture of what it takes to embed digital tools successfully into a national healthcare system.

Digital Pillar: A major driver of Japan's current reforms is its uniquely super-aged population: nearly 30 percent of Japanese citizens are 65 or older, the highest global proportion. This demographic shift puts substantial strain on physical healthcare resources such as hospitals, clinics, and frontline providers. It also creates regional imbalances. Large numbers of doctors practice in dense urban areas, leaving many rural or remote districts underserved. Although digital interventions such as telemedicine and remote monitoring could ease these pressures, physical infrastructure remains uneven. Broadband networks, hospital hardware, and reliable transport for lab samples or follow-up remain essential. No matter how advanced Japan's software becomes, these real-world constraints can limit the reach of digital care.

Even in a tech-savvy country, implementing digital health hinges on four interconnected components: *Interoperability, Infrastructure, Clinical Buy-in*, and *Workforce Training*. Japan lacks a universal EHR system; instead, hospitals often house their own isolated records, making data sharing laborious.[2] Efforts to pilot telehealth or remote monitoring tools can lead to incompatible technologies that do not talk to each other. Meanwhile, many providers who already grapple with busy schedules find it difficult to incorporate AI-based workflows or advanced analytics into daily practice. Converting enthusiasm into day-to-day reliance requires workforce training that goes beyond superficial tutorials. The government's modernization drive aims to close these gaps, but a complete overhaul of hospital IT systems and staffing approaches is neither quick nor simple.

Financial Pillar: Japan's Ministry of Health, Labour, and Welfare launched the Medical DX Promotion Plan in 2022 to modernize EHRs and encourage AI-driven diagnostics, but long-standing regulatory

---

[2] U.S. Chamber of Commerce. *Digitization Delivers: Japan's Digital Health Transformation*. Published 2024. Accessed February 11, 2025. https://www.usch amber.com/assets/documents/Digitization-Delivers-Japans-Digital-Health-Transformation.pdf.

frameworks often cause new digital tools to languish in year-long approval processes.[3] Even after initial clearance, updating AI systems or telehealth platforms can be delayed by bureaucratic inertia. Sustainable reimbursement presents another challenge. While the government broadened telemedicine payment policies during the COVID-19 crisis, the overall payment structure continues to favor in-person visits. AI-based diagnostics face particular scrutiny in national insurance coverage; developers must demonstrate both safety and cost-effectiveness through a complex, multi-step evaluation that slows regulatory alignment.[4] Only when these financial pieces work in tandem—leading to swift but thorough regulation and consistent reimbursement—can innovation take root.

Cultural Pillar: Japan's healthcare culture has traditionally revolved around face-to-face consultations, with older generations especially loyal to in-person care. This preference is not merely a habit, it stems from long-held trust in direct interaction with physicians. To change that outlook, health authorities have begun community engagement efforts such as public campaigns and pilot programs that introduce telemedicine as a convenient, reliable option rather than a drastic departure from the norm. Nonetheless, cultural hesitations remain. If a rural health clinic offers a video visit, but local residents perceive it as subpar compared to seeing a doctor in person, uptake may stall. True success thus hinges on respecting cultural values while illustrating the tangible benefits of remote and AI-assisted care. Government-led awareness campaigns have begun, but widespread digital literacy remains a challenge.

Trust/Knowledge Pillar: The final layer, *digital literacy*, directly influences both clinicians and patients. Many nurses, physicians, and hospital administrators remain skeptical about AI's clinical accuracy, fearing bias or unclear decision-making processes. Patients, particularly older ones, may be wary of transmitting personal health data through digital channels or relying on machine-driven diagnoses. Government-backed initiatives now offer educational sessions on privacy, data protection, and the basics of AI algorithms. Yet true trust is cultivated over time. Japan's path forward relies on demonstrating that new technologies can reinforce, rather than replace, the human touch at the core of healthcare.

Japan's attempt to modernize its healthcare delivery illustrates how all five pillars must align if large-scale change is to take root

---

[3] Ibid.
[4] Ibid.

(see Diagram 3.5). Even a highly developed country can struggle when entrenched cultural norms, fragmented EHRs, and protracted regulatory approvals converge. Although the Medical DX Plan has begun to address many of these issues, each pillar remains critical. The nation's progress highlights the importance of tackling these challenges in unison, showing that a truly cohesive digital health strategy depends not just on cutting-edge technology, but also on comprehensive support at every level of the healthcare system.

PILLARS	JAPAN'S SITUATION	IMPACT
Financial Pillar Regulatory Alignment	Lengthy AI-device approvals; extensive bureaucratic oversight slowing updates.	Innovation cycles dragged out, preventing quick deployment of new digital health tools.
Sustainable Reimbursement	Telemedicine payments added during COVID-19, but still favor in-person care; new AI diagnostics often lack clear, long-term funding.	Cost-effectiveness hurdles for AI solutions; limited telehealth growth beyond early pandemic surge.
Digital Pillar Interoperability & Infrastructure	Hospital-based record silos; minimal data-sharing standards.	Fragmented health data, making integrated or nationwide remote care solutions difficult to scale.
Clinical Buy-In & Workforce Training	Overburdened clinicians; skepticism about AI accuracy; insufficient training on digital workflows.	Cautious adoption; some providers doubt AI's reliability or dislike extra workload.

(continues)

(Continued)

PILLARS	JAPAN'S SITUATION	IMPACT
  **Cultural Pillar** Community Engagement 	Patient culture strongly values face-to-face visits; older adults reluctant to trust AI or remote monitoring; limited awareness campaigns.	Slower acceptance of telehealth and AI; government ramping up digital literacy efforts.
Trust/Knowledge Pillar Digital Literacy 		

Diagram 3-5: Japan's Five Pillars.

Singapore: AI-Driven Health in a High-Tech Nation

Singapore's reputation as a global innovation hub makes it an ideal testbed for AI-driven healthcare. Yet even in this compact, technologically advanced city-state, success rests on effectively blending the five pillars—addressing local physical realities, ensuring sound financial structures, engaging communities culturally, building robust digital systems, and fostering the trust and knowledge needed for widespread adoption.

Physical Pillar: Though Singapore does not grapple with vast rural territories, it confronts physical constraints in the form of hospital capacity limits and a busy urban environment. An efficient transport network and centralized hospital system enable quick access to specialized services, but overcrowding remains a concern at peak times. AI-based analytics help predict admission surges and reroute non-urgent cases, easing pressure on facilities.[5] Even in a high-density urban setting, digital initiatives cannot thrive unless they address very real challenges, such as bed space, patient flow, and clinical workforce distribution.

---

[5] Ibid.

Financial Pillar: A dual public-private financing model underpins Singapore's healthcare system. Programs like MediShield Life offer universal coverage for major hospital bills, while private insurance can supplement more specialized treatments. The Ministry of Health strengthens these mechanisms through clear regulatory alignment, issuing guidelines that reduce legal uncertainties around AI pilot projects. At the same time, sustainable reimbursement emerges as a key enabler. By co-funding promising digital initiatives, such as predictive analytics platforms, Singapore ensures that AI deployments can scale beyond brief, one-off trials. This alignment between clear rules and reliable funding helps innovators remain confident in the long-term viability of their tools.

Cultural Pillar: Despite Singapore's modern image, an aging population can be cautious about relying on new technologies. Government-led community engagement programs, part of the broader "Smart Nation" campaign, focus on presenting tangible benefits such as faster diagnoses, reduced crowding, and more personalized therapies. These efforts align digital solutions with societal expectations, showing that AI-driven health is not merely a novelty but a practical means of improving care quality. By framing AI as a complement to the existing doctor–patient relationship, rather than a replacement, public buy-in grows among those who might otherwise fear the unfamiliar.

Digital Pillar: A robust digital ecosystem is the key to making AI efforts truly effective. Interoperability—the seamless exchange of data among hospitals, clinics, and other providers—remains a work in progress, especially when private facilities use different EHRs. Still, Singapore benefits from strong infrastructure, notably extensive broadband coverage, and advanced IT platforms in major hospitals. Within these institutions, clinical buy-in often begins with small pilot projects run by "innovation teams" that partner with frontline clinicians.[6] When nurses and physicians witness real-world advantages, such as shorter wait times or quicker lab turnaround, they become more amenable to integrating AI into everyday practice. Dedicated workforce training then ensures that staff can interpret AI outputs responsibly and adapt workflows without undue disruption.

Trust/Knowledge Pillar: Even with supportive leadership and funding, digital healthcare hinges on building digital literacy among both medical professionals and the public. Many clinicians are initially

---

[6] Smart Nation Singapore. *Health Initiatives*. Accessed February 16, 2025. https://www.smartnation.gov.sg/initiatives/health.

skeptical about algorithms they cannot readily explain, and patients may worry about data security or biased decision-making. Singapore's authorities have responded by running workshops and public forums that detail how AI safeguards patient information and enhances care rather than supplanting human judgment. This steady, transparent communication has gone a long way toward building confidence in AI-enabled services; vital in a society that prizes reliability and proven results.

Singapore's high-tech ambitions rest on all five pillars working in tandem. **Physical Pillar** realities dictate where and how AI solutions relieve overcrowded hospitals. **Financial Pillar** systems ensure that new tools remain both affordable and regulatory-compliant. **Cultural Pillar** engagement fosters community trust in unfamiliar technologies. **Digital Pillar** underpinnings, especially interoperability, turn data into actionable insights. And ongoing training keeps clinical teams at ease with algorithmic tools. Finally, widespread digital literacy cements long-term acceptance, as patients and providers gain confidence in AI's capacity to deliver safe, impactful care. Although challenges remain, particularly around unifying private and public health records, Singapore's case illustrates how a forward-thinking, top-down approach can align policy, funding, and social outreach to make AI-driven care a reality at a national scale. It also shows that technology alone is not enough.

Technology alone isn't enough.

Digital Health Success Stories

For years, digital health has been heralded as the solution to healthcare gaps, yet many initiatives have failed to live up to their promise. Some crumble under poor infrastructure, others lack policy support, and many simply run out of funding before they can reach scale. But in some corners of the world, countries are proving that with the right strategy, digital health can become an integral part of national healthcare systems.

While it may seem like there are plenty of failures out there, there are also great examples of success. Kenya, India, and Rwanda have emerged as global leaders in sustainable digital health models, each leveraging technology, government support, and innovative financing to create lasting, impactful healthcare solutions. Their successes tell a different story than the digital health failures explored earlier—one of adaptation, scale, and long-term viability.

Kenya: A Mobile-First Healthcare Transformation

Kenya stands out among low- and middle-income countries for weaving digital health tools into its broader healthcare fabric, rather than confining them to small, donor-led pilots. Often cited for its dynamic mobile money ecosystem, the country has leveraged smartphone platforms, like M-TIBA and M-Pesa, to reduce financial barriers and expand care access. Yet Kenya's progress goes deeper than technology alone; it reflects a deliberate effort to align **Financial Pillar** strategies, **Digital Pillar** infrastructure, **Cultural Pillar** acceptance, and **Trust/ Knowledge Pillar** issues within the country's **Physical Pillar** realities.

Over the past decade, Kenya's Ministry of Health has refined its national eHealth policy to foster regulatory alignment, removing barriers that would otherwise hamper telehealth and AI ventures. This flexible environment encourages mHealth (mobile health) platforms to scale well beyond pilot status. Meanwhile, sustainable reimbursement models help ensure they can thrive in the long run. For instance, M-TIBA, a mobile-based health wallet, reduces out-of-pocket expenses by letting patients pay over time, integrating seamlessly with major insurers and providers.[7] In a system where financial stability is crucial, public-private partnerships (PPPs) further subsidize telehealth and AI-based services, making them affordable to low-income communities while preserving profitability for healthcare entrepreneurs.

Kenya's strong mobile connectivity in urban and *peri-urban* (a transitional zone located between urban and rural areas and tends to have characteristics of both) areas reflects its keen focus on **Physical Pillar** and **Digital Pillar** readiness.

Firms like Ilara Health deploy AI-powered diagnostic devices that transmit patient results to cloud-based systems, bypassing the need for large lab setups.[8] By standardizing data formats across clinics, these initiatives move the needle on interoperability, a linchpin of the **Digital Pillar**. Rural connectivity gaps remain a challenge, but Kenya's success with mobile money (M-Pesa) shows that large-scale digital adoption is achievable when government support and local user demand align.

---

[7] M-TIBA. *About Us: Accessible, Affordable Healthcare*. M-TIBA website. Accessed February 16, 2025. `https://mtiba.com/about-us-accessible-affordable-healthcare/`.

[8] Ilara Health. *HMIS—Hospital Management Information System*. Accessed February 16, 2025. `https://www.ilarahealth.com/hmis`.

On the ground, clinical buy-in and workforce training prove critical to any digital health rollout. Community health workers (CHWs) in Kenya are given digital devices and practical guidance, not only for collecting patient data but also for facilitating appointments and follow-ups.[9] The fact that doctors and nurses see faster test results, improved maternal care, and better chronic disease management spurs broader acceptance. However, scaling this effort nationwide is no small feat; thousands of clinics still need continuous IT support and refresher programs. Ensuring that **Physical Pillar** and **Digital Pillar** infrastructures meet high demand is central to keeping frontline staff enthusiastic, rather than overwhelmed.

Next Shift Quick Wins: Don't Launch Without a Fall-Back Plan

What to do before you deploy:

1. Write a plain-language one-pager titled: "What to do when this breaks."
2. Assign a person, not a platform, to respond to failures and be accountable.
3. Print the paper form version and run a 10-minute drill using it.

Kenya's **Cultural Pillar** context is a mosaic of urban centers, rural villages, and numerous local traditions. Consequently, community engagement becomes integral to each new mHealth platform's rollout, ensuring that telehealth and AI aren't perceived as foreign interventions. Government-led initiatives and NGO (Non-Government Organization) outreach explain how digital health solutions improve patient outcomes, while CHWs tailor this information to local languages and customs.[10] By fostering digital literacy, Kenya empowers families to navigate health wallets and AI-driven apps without fear of hidden costs or privacy breaches. Because let's be real; if people don't trust an app, they're not using it. (Just ask anyone who still won't put their credit card into a website they don't recognize.) This grassroots

[9] UNICEF Kenya. *Health*. Accessed February 16, 2025. https://www.unicef.org/kenya/health.

[10] World Health Organization (WHO). *Digital Health Literacy Key to Overcoming Barriers for Health Workers*. Published 2023. Accessed March 2, 2025. https://www.who.int/news/item/digital-health-literacy-key-to-overcoming-barriers-for-health-workers.

trust forms the backbone of community buy-in, reinforcing the notion that digital solutions can genuinely solve day-to-day problems.

Kenya's trajectory highlights the synergy among its **Physical Pillar** infrastructure, mechanisms, **Financial Pillar** and **Cultural Pillar** acceptance, **Digital Pillar** maturity, and **Trust/Knowledge Pillar** building (see Diagram 3.6). Mobile payment solutions, AI-based diagnostics, and community outreach have propelled the nation into a global spotlight. The country's success, however, is not just about inventive apps or advanced algorithms. It reflects a coordinated push—from governmental policies to provider incentives, to local champions—that ensures technology solutions address actual needs. While interoperability gaps and rural connectivity shortfalls persist, Kenya's example shows that mobile-first innovations can reshape healthcare at scale when financial stability, regulatory clarity, and local trust converge under a cohesive national framework.

PILLAR	KENYA'S SITUATION	IMPACT
Financial Pillar Regulatory Alignment	eHealth policy and Ministry of Health oversight encourage large-scale mobile health initiatives.	Reduced donor-driven fragmentation; clear rules for registering mHealth startups.
Sustainable Reimbursement	M-TIBA wallet plus public-private insurance collaborations; partial subsidies for digital health.	Expanded access for low-income patients; telehealth solutions remain affordable beyond pilots.
Digital Pillar Interoperability & Infrastructure	AI-driven devices (Ilara Health), mobile coverage strong in most urban areas; rural network gaps persist.	Clinics can leverage cloud-based diagnostics, though remote areas still face connectivity barriers.

(continues)

(Continued)

PILLAR	KENYA'S SITUATION	IMPACT
Clinical Buy-In & Workforce Training 	CHWs and clinic staff trained on digital platforms; CHWs collect data, provide basic counseling.	Growing acceptance of AI/telehealth among providers; consistent training needed at rural sites.
  **Cultural Pillar** Community Engagement	Grassroots outreach to families on mobile payment, telehealth; CHWs serve as trusted intermediaries.	Improved patient trust and use of digital tools; fosters local ownership of health programs.
Trust/Knowledge Pillar Digital Literacy		

Diagram 3-6: Kenya's Five Pillars.

Fact Check: Kenya by the Numbers

1. In 2020, over 88 percent of children were fully vaccinated, up from 84 percent in 2014.

2. Funding for vaccine procurement increased by 80 percent in 2020.

3. Under age five mortalities dropped by 57 percent between 1990 and 2019.

Across Africa: Contrasting Successes and Common Failure Points

Across Africa, nations are harnessing mobile technology to tackle widespread barriers in healthcare access. Rwanda shines as a leading example, demonstrating how well-aligned government policies and effective public-private partnerships can push digital health beyond small-scale pilots.[11] By contrast, many other African countries struggle to maintain similar initiatives once donor support ebbs or infrastructure issues resurface. Together, these experiences reveal not just flashes of innovation, but also the structural and cultural obstacles that so often undermine long-term viability.

Stories from the Field: Digital Pillar Spotlight in Rwanda

One of Rwanda's most celebrated achievements is its drone delivery network, operated by Zipline, which transports essential medical supplies, particularly blood and vaccines, to remote clinics at unprecedented speed.[12] Unlike many conventional quadcopter drones, Zipline's fixed wing "Zips" can traverse up to 150 kilometers on a single charge, handling tough weather while dropping payloads via parachute for rapid, hands-free retrieval.[13] Think of this like many of the commercial drone programs you've seen from Amazon to food delivery. Except instead of delivering diaper cream and tacos, it's plasma and whole blood. Since December 2016, Zipline has shipped over 4,000 units of blood products to a dozen hospitals, slashing delivery times from hours to mere minutes.[14] This feat underscores how strong government endorsement, combined with practical technology, can radically boost rural access, offering a living model of effective public-private collaboration.

---

[11] Africa: Rwanda's success in mobile-first health while others failed. https://www.medicaldevice-network.com/features/setting-an-example-rwanda-as-a-digital-health-success-story.

[12] Baker A. In Rwanda, drones deliver vital medical supplies, revolutionizing healthcare access. *TIME*. Published February 8, 2024. Accessed February 14, 2025. https://time.com/rwanda-drones-zipline.

[13] Ibid.

[14] Baker A. In Rwanda, drones deliver vital medical supplies, revolutionizing healthcare access. *TIME*. Published February 8, 2024. Accessed February 14, 2025. https://time.com/rwanda-drones-zipline.

Ghana

Elsewhere, the story is more complicated. In Ghana, a national telemedicine initiative funded by the Novartis Foundation set up nurse-run call centers, allowing rural patients to get expert triage and remote doctor consults.[15] Although this approach initially improved emergency response times, it floundered as external funding waned, and local healthcare structures failed to integrate it fully. By the time COVID-19 struck, the program's relevance was crystal clear, yet it still faced inconsistent technology adoption, mounting staff shortages, and an absence of robust, government-led financing.[16] What had begun as a promising model dissolved under the weight of insufficient long-term planning and limited policy support **(Financial Barriers)**.

Senegal and South Africa

Similar narratives emerge in Senegal and South Africa, where text-based interventions were launched to reduce maternal mortality. Early pilot data seemed encouraging, until poor Internet connectivity, low smartphone penetration, and **Cultural Barrier** skepticism scuttled large-scale rollout. Many pregnant women simply did not trust **(Trust/Knowledge Barrier)** or prioritize an SMS reminder over traditional face-to-face advice. Without lasting state sponsorship or a stable reimbursement framework, these digital platforms remained stuck in a perpetual pilot phase.[17]

Bangladesh

This pattern also appears beyond Africa, as seen in Bangladesh's Mobile Alliance for Maternal Action (MAMA), designed to enhance prenatal care compliance via SMS.[14] While it gained momentum early on, many users dropped off due to cellular costs, uneven coverage, or

---

[15] Novartis Foundation and Ghana Health Service announce successful integration and scale-up of telemedicine program. Novartis Foundation. Published July 9, 2018. Accessed February 14, 2025. https://novartisfoundation.africa-newsroom.com/press/novartis-foundation-and-ghana-health-service-announce-successful-integration-and-scaleup-of-telemedicine-program?lang=en.

[16] Dzando G, Akpeke H, Kumah A, et al. Telemedicine in Ghana: insight into the past and present, a narrative review of literature amidst the Coronavirus pandemic. *J Public Health Afr.* 2022;13(1):2024. doi:10.4081/jphia.2022.2024.

[17] All digital health scale-up efforts should be accompanied by snacks, shade, and an emotionally available IT liaison who can explain new APIs without sighing.

lingering doubts about whether a text could replace human contact. A systematic review of SMS-based health campaigns found that illiteracy and minimal *digital literacy* hampered communication, especially among populations unable to read or interpret text messages.

Despite pockets of success, many telemedicine and mHealth projects in Africa and similar contexts collapse after promising pilots. Often, they operate outside formal health structures, relying on donor funds with minimal government integration. Once external money dissipates, local systems struggle to sustain them. This is particularly true if national policies and reimbursement models never fully absorbed the technology, enforcing a **Financial Barrier**. Even well-designed cross-border telehealth ventures falter when local licensing rules and data protection laws remain unclear, blocking expansion into remote districts.

In addition, regions with frequent power outages, slow Internet, or limited smartphone usage cannot fully realize the benefits of mHealth, no matter how compelling the innovation. Where connectivity does exist, many communities hesitate to trust **(Trust/Knowledge Barrier)** remote consultations, perceiving app- or SMS-based care as inferior to in-person visits. Without focused efforts to win acceptance **(Cultural Barrier)**, build frontline capacity, and adapt solutions to varying literacy levels, digital health projects struggle to gain user loyalty. These recurring setbacks remind us that technology alone does not suffice. If mHealth is to transcend pilot status, it must become part of broader national strategies, covered by workable funding and embraced by the very communities it aims to serve.

Nevertheless, Rwanda's Zipline experience shows that targeted government support and innovative engineering can conquer even the toughest logistical hurdles. For other African nations, bridging the gap between bright pilot ideas and large-scale rollouts means addressing gaps early, forging robust partnerships, and embedding programs into existing health systems so they do not wither once donors exit. Meanwhile, Ghana's stumbling telemedicine initiative and Senegal's SMS maternal health project illustrate that scaling demands more than tech savvy, it calls for long-term financing **(Financial Pillar)**, dedicated training, and public trust **(Trust/Knowledge Pillar)**.[18] Only by tackling infrastructure shortfalls, aligning with national healthcare priorities, and fostering local acceptance can Africa's mobile health revolution move from sporadic promise to enduring, life-saving impact.

---

[18] Maternal and Child Survival Program. *Mobile Alliance for Maternal Action (MAMA) Lessons Learned.* Published April 2018. Accessed February 14, 2025. https://mcsprogram.org/resource/mama-lessons-learned-report.

Stories from the Field: Digital and Financial Pillars Spotlight in India

With over 1.4 billion citizens, India faces monumental healthcare challenges, from remote rural villages to overcrowded urban hospitals. Yet over the past decade, the country has embraced a system-wide approach to filling gaps and addressing needs rather than relying on fragmented pilots. Government platforms like eSanjeevani and grassroots solutions like Kilkari have accelerated telemedicine adoption by integrating with national health frameworks, improving connectivity, and earning the trust **(Trust/Knowledge Pillar)** of local communities.

India's model for digital health begins with regulatory alignment. Recent National Telemedicine Guidelines clarify licensing and patient consent, offering providers the legal certainty to conduct remote consultations. At the same time, the National Digital Health Mission (NDHM) aims to unify data standards **(Digital Pillar)** and create a single health ID, streamlining everything from insurance claims to referrals.[19]

On the financial side **(Financial Pillar)**, publicly funded telemedicine via eSanjeevani underscores India's belief in sustainable reimbursement. By offering virtual consultations free or at low cost, government channels enable rural and low-income families to access care they might otherwise skip.[20] Private insurers are beginning to reimburse teleconsultations too, though coverage can vary widely because India's insurance landscape remains a patchwork of different payers.

India's large and uneven terrain spotlights ongoing **Physical Barrier** challenges, from rural broadband gaps to overstretched clinics. Nevertheless, eSanjeevani provides a common platform **(Digital Pillar)**, standardizing telemedicine workflows across states. Since its launch, it has served over 340 million patients and engaged over 230,000 providers, reducing the burden on big-city hospitals while sparing patients' lengthy travel.[21] In regions without reliable broadband, programs pivot to low-bandwidth solutions, such as SMS-based

---

[19] Kruse CS, Stein A, Thomas H, Kaur H. Realizing the potential of telemedicine in global health. *BMC Med*. 2018;16(1):214. Accessed March 2, 2025. https://www.ncbi.nlm.nih.gov/pmc/articles/PMC6281195.

[20] Ministry of Health and Family Welfare, Government of India. *eSanjeevani—About*. Accessed February 16, 2025. https://esanjeevani.mohfw.gov.in/#/about.

[21] World Economic Forum. *How Data-Driven Digital Tools Cut Costs and Boost Outcomes*. Published 2022. Accessed March 2, 2025. https://www.weforum.org/reports.

alerts or voice calls, adapting the technology to local realities instead of waiting for a perfect infrastructure.[22]

In a country with finite medical staff for a huge population, clinical buy-in and ongoing workforce training are critical. Government mandates encourage hospitals and primary health centers to adopt telehealth as an "official" tool, helping doctors see it as more than a side project. For frontline workers, short online courses and tablet-based triage apps offer a practical introduction to digital platforms. Yet the same overextended clinicians, who treat hundreds of patients daily, require support to truly integrate new workflows. Scaling telehealth in India thus hinges on bridging the expertise gap **(Digital Barrier)** without further overwhelming providers.

India's diversity is both a challenge and an opportunity for digital health. Community engagement, often through well-known government brands, builds trust **(Trust/Knowledge Pillar)** among populations of varying language and literacy levels. For instance, Aarogya Setu, originally a COVID-19 contact-tracing app, expanded to include telehealth services, reaching millions of Indians within weeks of launch.[23] Meanwhile, Kilkari, an audio-based maternal health program available in eight local languages, exemplifies digital literacy tailored to a context where reading barriers might hinder SMS use.[24] By communicating key health messages through voice calls, Kilkari reaches mothers who might otherwise miss vital prenatal or postnatal care information.

India's massive scale and complex health system might seem daunting, but the country's steady progress shows what happens when physical infrastructure, financial backing **(Financial Pillar)**, outreach **(Cultural Pillar)**, digital innovation, and trust-building **(Trust/Knowledge Pillar)** converge (see Diagram 3.7). Telehealth guidelines remove legal roadblocks, public platforms like eSanjeevani reduce costs for underserved communities, and community-driven apps like Kilkari ensure that no region is left behind by technological progress. Still, work remains—extending broadband to rural districts, refining data exchange, and easing the workload on overburdened

---

[22] Telecom Regulatory Authority of India (TRAI). *TRAI Annual Report 2024.* Published January 2025.

[23] Suresh V. Aarogya Setu – Digitized COVID-19 contact tracing. *Disaster Med Public Health Prep.* 2022;16(4):1522–1525. Accessed February 16, 2025. https://pmc.ncbi.nlm.nih.gov/articles/PMC9148629/pdf/DIMD90000000.0-00007.pdf.

[24] ARMMAN. *Kilkari.* Accessed February 16, 2025. https://armman.org/kilkari.

medical staff. Yet India's example proves that even the world's largest populations can adopt telehealth, AI, and mobile health solutions when policy, funding, infrastructure, training, and local engagement strengthen each other in turn.

PILLAR	INDIA'S SITUATION	IMPACT
Financial Pillar Regulatory Alignment	National Telemedicine Guidelines, NDHM create clear legal pathways for virtual care.	Eases provider concerns about licensing; fosters nationwide standardization.
Sustainable Reimbursement	eSanjeevani is publicly funded; some private insurers reimburse telehealth.	Low-income groups have subsidized access; private coverage remains inconsistent.
Digital Pillar Interoperability & Infrastructure	eSanjeevani serves as a shared telehealth platform; rural broadband is still patchy.	Unified approach to data exchange in government settings; connectivity challenges persist.
Clinical Buy-In & Workforce Training	Government mandate for public providers to integrate telehealth; short courses train clinicians on virtual consults.	Official endorsement increases usage; ongoing training required to maintain adoption.

PILLAR	INDIA'S SITUATION	IMPACT
Cultural Pillar Community Engagement 	Aarogya Setu and Kilkari use mobile apps and voice messaging to reach diverse populations, including nonliterate users.	High uptake in rural and underserved areas; culturally adapted outreach sustains momentum.
Trust/Knowledge Pillar Digital Literacy 		

Diagram 3-7: India's Five Pillars.

Stories from the Field: Financial and Digital Pillars Spotlight in Estonia

Estonia's reputation as an e-governance frontrunner stems from deliberate investments in **Physical Pillar** readiness, **Financial Pillar** support, **Cultural Pillar** acceptance, **Digital Pillar** systems, and **Trust/ Knowledge Pillar** building. Over the last few decades, the country has enacted strong regulatory alignment, ensuring that electronic signatures, digital records, and patient data protection are legally recognized. This clarity underpins the nation's entire digital infrastructure, letting healthcare providers and patients interact within a unified, tech-driven environment.

From the start, Estonia folded sustainable reimbursement into its national health insurance system, treating digital services, such as e-prescriptions and teleconsultations, as part of standard care. Because these costs are directly built into regular reimbursement structures, digital platforms do not depend on scattered pilot funding. Meanwhile, the government's forward-looking regulatory alignment grants legal backing to mHealth innovations, from streamlined licensing for telemedicine services to detailed data protection statutes.[25] Together, these financial **(Financial Pillar)** and policy measures create a stable environment in which healthcare tech can thrive.

On the **Physical Pillar** side, Estonia is relatively compact, but the country still needed robust connectivity to link urban hospitals with outlying clinics. The result is near-ubiquitous broadband coverage and national support for a digital backbone that unites all healthcare players. The entire citizenry is connected to a single EHR system, reflecting a commitment to interoperability and infrastructure, so that data flows securely across clinics, pharmacies, and hospitals.[26] Paper-based prescriptions are a rarity. Digital scripts eliminate errors and cut patient wait times, demonstrating how thoughtfully designed technology can resolve logistical snags in day-to-day care.

Next Shift Quick Wins: Put Interop on the Wall

No one cares about backend plumbing, until it leaks.

- Post a real-time API uptime poster in the nurses' lounge.
- Translate "FHIR compliant" into "Can your notes follow the patient?"
- Celebrate zero-error weeks with snacks (and optional bragging rights).

Estonia's success also owes much to clinical buy-in and workforce training. From day one, healthcare professionals have been included in system design and given practical, standardized training in EHR usage and privacy protocols. Because the digital platform is official and pervasive, clinicians see direct advantages: faster documentation, fewer administrative hurdles, and clearer patient histories. This user-friendly approach helps normalize technology in the clinical workspace, transforming it from an occasional experiment into the default method of managing records and prescriptions.

Equally important is community engagement and digital literacy, which fortifies the **Trust/Knowledge Pillar** among the public. Estonia's transparent data-access policies let citizens view exactly who has

[25] e-Estonia. *Healthcare Is Changing, and Technology Is Changing with It.* Accessed February 16, 2025. https://e-estonia.com/healthcare-is-changing-and-technology-is-changing-with-it/.
[26] Ibid.

opened their health records.[27] This openness, combined with a populace well-versed in e-services, breeds confidence in the system's security. Rather than fearing misuse of private information, most patients trust the mHealth framework, and many expect digital convenience as a baseline. In an environment where people routinely sign documents online and conduct official business digitally, mHealth appears as a natural extension of daily life, not an external imposition.

Estonia's distinction as a digital health pioneer arises from uniting its physical infrastructure, a **Financial Pillar** model, **Cultural Pillar** readiness, a digital backbone, and **Trust/Knowledge Pillar** initiatives into one seamless system. Regulatory alignment ensures that mHealth tools operate within clear legal frameworks, while sustainable reimbursement keeps them fiscally viable long term. Interoperability and infrastructure link providers and patients effortlessly, and clinical buy-in and workforce training cement technology's place in routine practice. Finally, community engagement and digital literacy instill broad public confidence, making mHealth services ubiquitous rather than experimental. In Estonia, digital health is not an add-on; it is the default mode of care, rooted in a nationally integrated framework that other nations aspire to replicate.

Making Digital Health Stick: Pulling It All Together

You have seen throughout this chapter that no single factor guarantees success when it comes to digital health. Even the most promising telehealth app, AI-based diagnostic, or remote monitoring device can falter if the larger ecosystem is unprepared. The **Physical Pillar**, **Financial Pillar**, **Cultural Pillar**, **Digital Pillar**, and **Trust/Knowledge Pillar** each play a role in creating the system-level readiness needed to nurture fledgling technologies and guide them into everyday practice. When all five pillars mesh with practical considerations, the stage is set for robust adoption:

- Physical gaps shrink when local advocates push for expansions or upgrades, while accessible financing grows stronger under supportive policies and reimbursement models.

---

[27] e-Estonia. *Healthcare Is Changing, and Technology Is Changing with It.* Accessed February 16, 2025. https://e-estonia.com/healthcare-is-changing-and-technology-is-changing-with-it/.

- Cultural acceptance flourishes when champions engage communities and iteration incorporates local voices.

- The digital ecosystem matures when training and agile development keep solutions fresh.

- Trust and knowledge deepen as transparent policies safeguard data privacy and familiar faces (whether they are clinicians or community leaders) guide new users through the technology.

In short, digital health truly "sticks" when synergy forms across each of these domains. Otherwise, it's just another high-tech fad collecting dust next to Google Glass and that one smart fridge nobody really needed.

Fact Check[28]

1. The percentage of physicians who feel digital health tools are an advantage for patient care grew from 85 percent in 2016 to 93 percent in 2022.

2. Improved clinical outcomes and work efficiency are the top factors influencing physician interest in digital health tools.

3. The digital health tools that garner the most enthusiasm among physicians are tele-visits (57%) followed by remote monitoring devices (53%).

Digital health doesn't fail because of bad code, it fails because of mismatched context. It fails when platforms get launched into systems that aren't ready to catch them; when algorithms are unveiled without training, funding, or trust. What you've seen across Uganda, Japan, Kenya, and beyond is this—success doesn't hinge on the flashiest AI or the most elegant interface. It hinges on systems—the policies, payment structures, infrastructure, and cultural readiness that determine whether a great idea becomes a real-world solution. To make digital health stick, we must stop asking "what can the tech do?" and start asking "what does the system need?" Only then can we move from pilots to permanence and finally transform potential into progress.

---

[28] American Medical Association. *AMA Digital Health Care 2022 Study Findings.* AMA Website. Published 2022. Accessed February 17, 2025. https://www.ama-assn.org/about/research/ama-digital-health-care-2022-study-findings.

🔧 Startup Builder's Box: From "Pilotitis" to Platform: Building Digital Health That Sticks

Your Challenge: You've built a brilliant AI health platform (and of course it solves a real and urgent problem!), but ministries are overwhelmed, providers are skeptical, and you're one of 57 pilots crowding the system. What now?

Startup Levers to Pull:

- **System Fit First:** Before scaling, ask: *Does our product align with national regulatory strategy, provider workflows, and financing models?*

- **Localize for Trust:** Partner with community health workers and regional champions to co-design a deployment strategy and messaging.

- **Use the Scorecard:** Rate yourself across the system-level pillars: Interoperability, Clinical Buy-in, Digital Literacy, Reimbursement, and Regulatory Alignment.

- **No Orphans:** Ensure your tech doesn't die post-grant/pilot. Secure commitments for sustained reimbursement and policy endorsement.

- **Train the Trainers:** Build toolkits for regional health systems to evaluate, onboard, and support new tech independently.

- **Never Pilot Alone:** Join consortia that align with national or regional health IT priorities, instead of adding one more digital orphan.

 Field-tested Insight: Success in India, Estonia, and Singapore shows it's not about how shiny the tech is, it's how deeply it's wired into the health system it aims to support. (*This is also true in the United States!*)

Leaders' End-of-Chapter Action Checklist: Chapter 3: "The Missing Link"

LEADER	HIGH-IMPACT ACTION TO STRENGTHEN SYSTEM-LEVEL DIGITAL HEALTH INTEGRATION
❑ Board Director	Mandate that all digital pilots align with national data standards; require quarterly reporting on interoperability and sustainability metrics
❑ Chief Executive Officer	Establish a Systems Innovation Task Force within 30 days; publish a 12-month roadmap that aligns tech rollouts with workforce and reimbursement readiness
❑ Chief Information Officer	Launch a live dashboard tracking data exchange between internal and external platforms; flag weekly integration errors and downtime
❑ Chief Health Information Officer	Decommission all noncompliant mHealth apps; require documented FHIR compatibility and security standards for continued access
❑ VP Clinical Operations	Conduct quarterly drills on digital tool breakdowns; build and test fallback workflows for AI, EHR, and telehealth systems
❑ VP Nursing and Patient Education	Run system-wide bootcamps on clinical documentation standards, AI safety, and digital triage; reach 85% nursing staff completion by year-end
❑ VP Data and Analytics	Audit five live systems for algorithm bias, missing data harms, and explainability; deliver recommendations at the Systems Task Force quarterly
❑ Telehealth Program Manager	Pair every virtual care launch with synchronous workflow mapping and provider feedback sessions; resolve major issues in 30 days
❑ Patient Experience Manager	Launch a "System Fit" feedback survey for patients and clinicians; target a 15% rise in confidence in digital care handoffs
❑ Community Health Worker Supervisor	Lead neighborhood mapping of public Wi-Fi, device access, and app usability; use results to prioritize rollout zones
❑ Director of Snacks and Morale	Build a life-size cardboard cutout of the national digital health roadmap; each milestone unlocked earns a commemorative snack and a dramatic reading of interoperability success stories

Virtual Care Unleashed

Picture this: A paramedic in rural Washington State pricks a fingertip and 30 seconds later a cardiologist in Seattle sees the troponin spike on their tablet. Then a flight crew lifts off before the patient even reaches the local clinic. That chain of pixels, algorithms, and human judgment is virtual care in action. It converts distance into a rounding error and time into a clinical asset. When designed around the Five Pillars of Healthcare Access, such encounters do not merely patch gaps; they rebuild the roadway that carries care to every doorstep, barn, church basement, and city bus stop. Fail to honor one pillar and the result is another shiny experiment. Sustain all five and virtual care becomes the new connective tissue of modern medicine (see Diagram 4.1).

Diagram 4-1: Pillar Spotlight.

Making Sense of the Vocabulary (Before We Dive In, What Are We Even Talking About!?)

Virtual Care: The umbrella term. Any remote interaction, synchronous or asynchronous, that shares clinical or supportive information between a patient and the circle of care through digital channels with the goal of improving health outcomes or the care experience. This spans telephone check-ins, secure app messaging, remote patient-monitoring dashboards, virtual reality rehabilitation, and even robotic-assisted surgery consoles.

Telehealth: A broad subset of virtual care that includes clinical services *and* non-clinical activities such as professional education, administrative meetings, and public-health outreach, delivered via information and communication technology. MyChart medication refills, medical-school grand rounds streamed on Zoom, and an insurer's video-based diabetes-prevention class all fit here.

Telemedicine: The clinical core. Real-time, two-way audiovisual encounters between a clinician at a distant site and a patient at an originating site, plus store-and-forward image review and e-consults that end in a diagnosis, a treatment plan, or a prescription. Think dermatology rash photos, post-operative video visits, and cardiology e-consults that shorten waitlists.

See Diagram 4.2 for a look at the relationship between these.

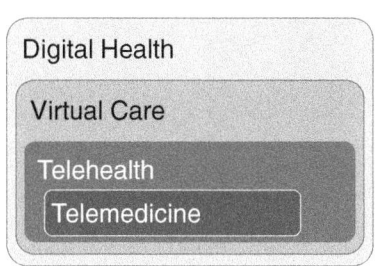

Digital health: Broad umbrella term encompassing eHealth as well as developing areas such as the use of advanced computing sciences (e.g., in the fields of big data and artificial intelligence)

Virtual care: Any interaction between patients and/or members of their circle of care that occurs remotely, using any forms of communication or information technology, with the aim of facilitating or maximizing the quality and effectiveness of patient care

Telehealth: The use of digital technologies to deliver medical care, health education, and public health services by connecting multiple users in separate locations

Telemedicine: A practice of medicine involving the use of information and communication technologies (ICT) by a health care provider to administer health care to patients

Diagram 4-2: Relationship Between Virtual Care, Telehealth, and Telemed.

So why does the taxonomy even matter? Policymakers write payment rules around these terms; technology teams map features to them and payers decide whether their insurance will pay. Clear language keeps every stakeholder in the same lane. Otherwise, muddled language breeds reimbursement denials, workflow confusion, and mistrust.

The Evolution of Virtual Care

Virtual care has transformed healthcare delivery by enhancing accessibility, reducing costs, and improving patient outcomes. This shift addresses longstanding challenges in healthcare access, particularly for remote and underserved populations. The COVID-19 pandemic further accelerated the adoption of virtual care, highlighting its critical role in maintaining healthcare continuity during crises.[1] We see

---

[1] Kichloo A, Albosta M, Dettloff K, et al. Telemedicine, the current COVID-19 pandemic and the future: a narrative review and perspectives moving forward in the USA. *Fam Med Community Health.* 2020;8(3):e000530.

telemedicine solving problems of access but also solving problems of patient and provider dissatisfaction. In fact, many providers consider their virtual experience with their patients just as good, if not better, than their office experience (see Diagram 4.3).

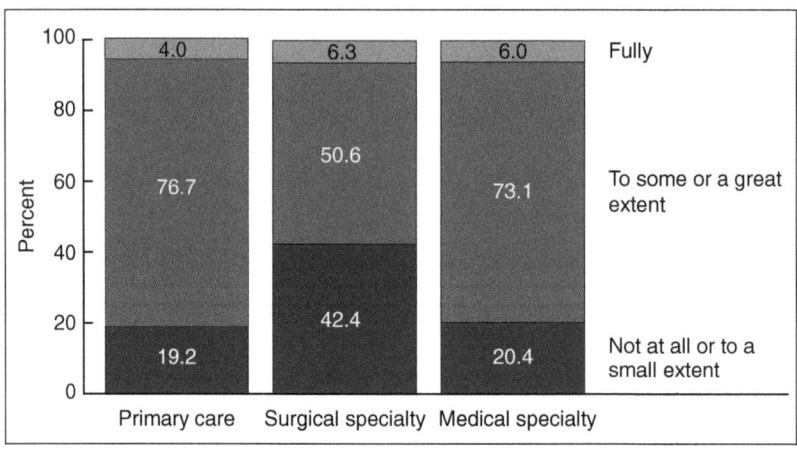

Diagram 4-3: Percentage of Physicians Able to Provide a Similar Quality of Care Utilizing Telemedicine Visits as Opposed to In-person. By Specialty, United States 2021.[2]

For my fire and EMS colleagues, COVID-19 forced a complete rethinking of emergency response. Many EMS teams adopted tele-triage protocols, where patients were assessed remotely before deciding whether an ambulance was necessary or if a patient truly needed to be transported. This not only protected healthcare workers but also optimized emergency response times. Telemedicine triage from the dispatch center has been used in England to safely treat low-acuity calls without needing in-person provider assessment.[3] EMS providers for a long time in the United States have utilized voice calls with their medical directors during times of a needed consultation, but the use of video has now been shown to be a positive addition for many pre-hospital indications. Tele-stroke services have been utilized across the United States. They have been used to diagnose stroke as early as possible, often by neurologists using telemed to an Emergency Department, or to EMS providers. In one study,

---

[2] National Center for Health Statistics. *Telemedicine Use Among Adults: United States, 2021*. NCHS Data Brief, no. 493. Published October 2023. Accessed March 11, 2025. https://www.cdc.gov/nchs/data/databriefs/db493.pdf.

[3] Haskins B, Brown KM, Evans J. Prehospital telemedicine: a systematic review of the literature. *J Telemed Telecare*. 2021;27(6):333–341. doi:10.1177/1357633X19892632.

the hospital network's median "door- to-needle" time decreased from 75 minutes in 2008 to 63 minutes in 2017.[4] This means that important *fibrolytics* ("clot busters") were administered much earlier, saving precious brain tissue.

While we expect a video call or an app today for our tele-visits, it didn't start that way. The modality of telemedicine has evolved significantly over time. The history of telehealth is an interesting one, both in regulations and policy and in the modality in which it is delivered (see Diagram 4.4).

History of Telehealth

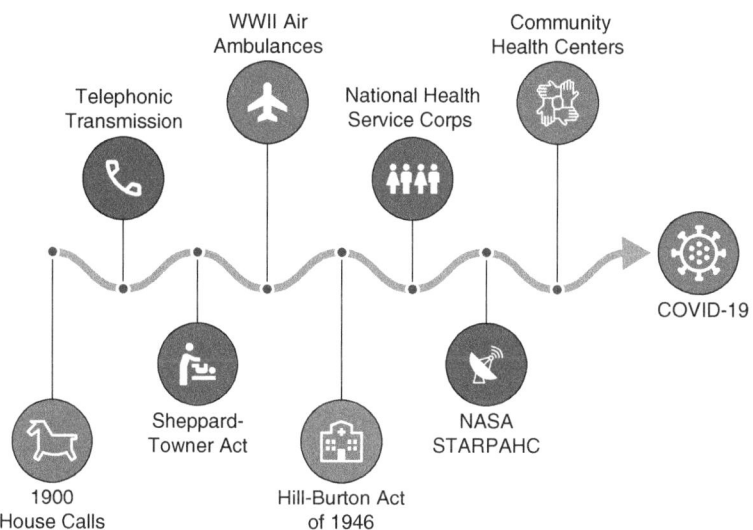

Diagram 4-4: Timeline of U.S. Healthcare Access Improvements and Telemedicine.

- **House Calls:** In the early 20th century, physicians commonly visited patients' homes, providing personalized care but facing limitations in reach and efficiency. My Austin house call practice mirrored this in many ways, but without the horse and buggy. I saw patients in a high touch fashion, but I could not see that many patients in a day.

---

[4] Winburn AS, Brixey JJ, Langabeer JR II, Champagne-Langabeer T. A systematic review of prehospital telehealth utilization. *J Telemed Telecare.* 2018;24(7):473–481. doi:10.1177/1357633X17713140.

- **Telephonic Transmission:** In 1905 the Dutch physician, Willem Einthoven, demonstrated the feasibility of telephonic transmission of heart sounds over a distance. Five years later, two American physicians successfully transmitted electrocardiograms of a variety of cardiac conditions across New York City.[5]

- **Sheppard-Towner Act (1921):** One of the earliest federal efforts to expand healthcare access, funding maternal and infant health programs that brought services to rural and underserved populations—precursor to modern outreach and remote care strategies.

- **Hill-Burton Act (1946):** Provided federal funding for hospital construction in underserved areas, laying the groundwork for infrastructure that would eventually support remote care and community-based services.

- **Early Telemedicine:** The 1960s marked the advent of telemedicine, with NASA pioneering remote medical monitoring for astronauts during space missions.[6] Concurrently, the University of Nebraska utilized two-way television to facilitate video consultations between clinicians and students.

- **COVID-19 Boom:** The global pandemic necessitated rapid adoption of telehealth services, enabling safe and effective patient care while minimizing infection risks.[6]

- **Telehealth Today:** Currently, telehealth integrates advanced technologies, including AI, to enhance diagnostic accuracy, personalize treatment plans, and streamline healthcare workflows. *If you have access to it.* Telehealth experiences are all over the spectrum on the technology used and the experience they provide.

While the concept of remote care is not new, technological advancements have significantly expanded its scope and effectiveness. The integration of AI and machine learning represents a novel development, enabling predictive analytics and decision support that were previously unattainable. These innovations differentiate contemporary virtual care from earlier models, offering more scalable and data-driven solutions to healthcare challenges (see Diagram 4.5).

---

[5] Nesbitt TS. The evolution of telehealth: where have we been and where are we going? In: *The Role of Telehealth in an Evolving Health Care Environment: Workshop Summary.* Board on Health Care Services; Institute of Medicine. National Academies Press (US); 2012. Accessed March 9, 2025. https://www.ncbi. nlm.nih.gov/books/NBK207141.

[6] American Hospital Association. *Fact Sheet: Telehealth.* Published February 7, 2025. Accessed March 6, 2025.

Despite technological progress **Digital Barriers**—such as digital literacy, broadband availability, and socioeconomic disparities—persist. Addressing these challenges requires ongoing efforts to ensure access to virtual care services, emphasizing the need for inclusive and adaptable healthcare strategies.

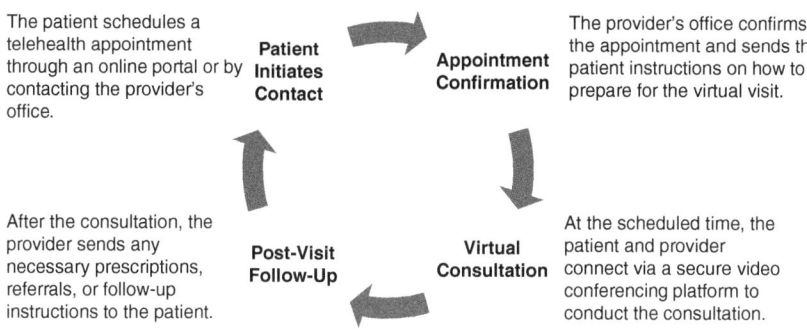

The patient schedules a telehealth appointment through an online portal or by contacting the provider's office.

Patient Initiates Contact

Appointment Confirmation

The provider's office confirms the appointment and sends the patient instructions on how to prepare for the virtual visit.

After the consultation, the provider sends any necessary prescriptions, referrals, or follow-up instructions to the patient.

Post-Visit Follow-Up

Virtual Consultation

At the scheduled time, the patient and provider connect via a secure video conferencing platform to conduct the consultation.

Diagram 4-5: Generalized Telehealth Process Flow.

Fact Check

1. During the COVID pandemic, Duke University created a virtual intensivist service (Tele-ICU) that allowed specialists to remotely manage intubated patients. Providers could see the ventilator settings and the patients' work of breathing, and remotely consult with the bedside team, all while decreasing exposure risk and preserving personal protective equipment (PPE).[7]

2. A study of over 35 million records by Epic (the EHR provider) found that for most telehealth visits across 33 specialties, there was no need for an in-person follow-up visit within 90 days of the telehealth visit.[8]

3. Having chronic conditions, multiple healthcare visits, and female sex increases the odds of having telemedicine visits. Older age, no Internet use, and living in the Midwest decrease the odds.[9]

---

[7] Kichloo A, Albosta M, Dettloff K, et al. Telemedicine, the current COVID-19 pandemic and the future: a narrative review and perspectives moving forward in the USA. *Fam Med Community Health.* 2020;8(3):e000530.

[8] National Center for Health Statistics. *Telemedicine Use Among Adults: United States, 2021.* NCHS Data Brief, no. 493. Published October 2023. Accessed March 11, 2025. https://www.cdc.gov/nchs/data/databriefs/db493.pdf.

[9] Chang E, Penfold RB, Berkman ND. Patient characteristics and telemedicine use in the US, 2022. *JAMA Netw Open.* 2024;7(3):e243354. doi:10.1001/jamanetworkopen.2024.3354.

Beyond the Screen: Making Virtual Care Truly Meaningful

While the technology continues to improve, as you've seen through the history of telehealth, there are always barriers and new things to consider. Virtual care represents more than simply transferring the traditional office visit to a digital platform; it embodies a fundamentally new approach to healthcare delivery. Much like the house calls of earlier medical eras, when doctors visited patients directly in their homes, virtual care now allows medical expertise to "enter" the patient's environment. But instead of knocking on their door, they enter through screens, breaking **Physical Barriers** and bringing healthcare into homes, workplaces, and communities instantly and seamlessly.

To truly transform healthcare, virtual care must not simply exist as a technological convenience. Instead, it needs to become seamlessly integrated into the everyday experiences of patients and providers. To achieve this, our understanding of healthcare access must evolve far beyond mere availability. It must also encompass the effectiveness, efficiency, and meaningfulness of everyday interaction, ensuring each virtual care experience genuinely meets patients' needs. In an ideal virtual healthcare environment, technology fades into the background, becoming so intuitive and unobtrusive that patients hardly notice its presence. When my teams were designing consumer/patient facing applications, we strove to give the users a "delightful" experience. I personally don't remember the last time as a patient I was "delighted" by any healthcare experience! Do you?

Imagine connecting effortlessly with healthcare providers, where each interaction feels as simple and natural as a face-to-face conversation. To reach this goal, we must eliminate common frustrations such as connectivity disruptions, overly complicated software, and cumbersome workflows. The future of healthcare access is one where technology seamlessly blends into daily life, ensuring that patients can effortlessly engage with their healthcare providers, fostering a connection as personal and immediate as an in-person conversation.

True continuity in healthcare emerges when virtual care is not treated as an isolated service but is deeply woven into the broader healthcare ecosystem. The most successful implementations enable smooth transitions between remote and in-person interactions, fostering comprehensive, continuous patient experiences. These systems excel because they seamlessly integrate virtual care into daily workflows, providing personalized and uninterrupted care across diverse settings. When providers have real-time access to the right information through interconnected technologies, they can make informed decisions when it matters most, greatly improving patient outcomes. The key to achieving continuity lies not only in the technology itself, but in creating a cohesive environment where digital solutions naturally complement and strengthen traditional healthcare (see Diagram 4.6). But to be successful there are many barriers to be overcome.

PILLAR	KEY FACTORS IN VIRTUAL CARE	EXAMPLE IMPACT ON VIRTUAL CARE SUCCESS
Physical Pillar 	Broadband, remote monitoring tools, accessible facilities	Without reliable broadband, telehealth sessions fail, limiting access for rural patients
Financial Pillar 	Reimbursement parity, sustainable funding	Without adequate reimbursement, providers can't sustainably offer telehealth
Cultural Pillar 	Cultural competence, local advocates, multilingual support	Lack of culturally competent telehealth can lead to patient mistrust or disengagement
Digital Pillar 	Digital literacy, interoperability, infrastructure reliability	Poor digital literacy reduces adoption and usability of telehealth systems
Trust/Knowledge Pillar 	Knowledge gaps on when and how to get in touch, confidence in secure communications	Without patient trust and education, adoption is slow, causing virtual care programs to stall

Diagram 4-6: The Five Pillars of Access Applied to Virtual Care.

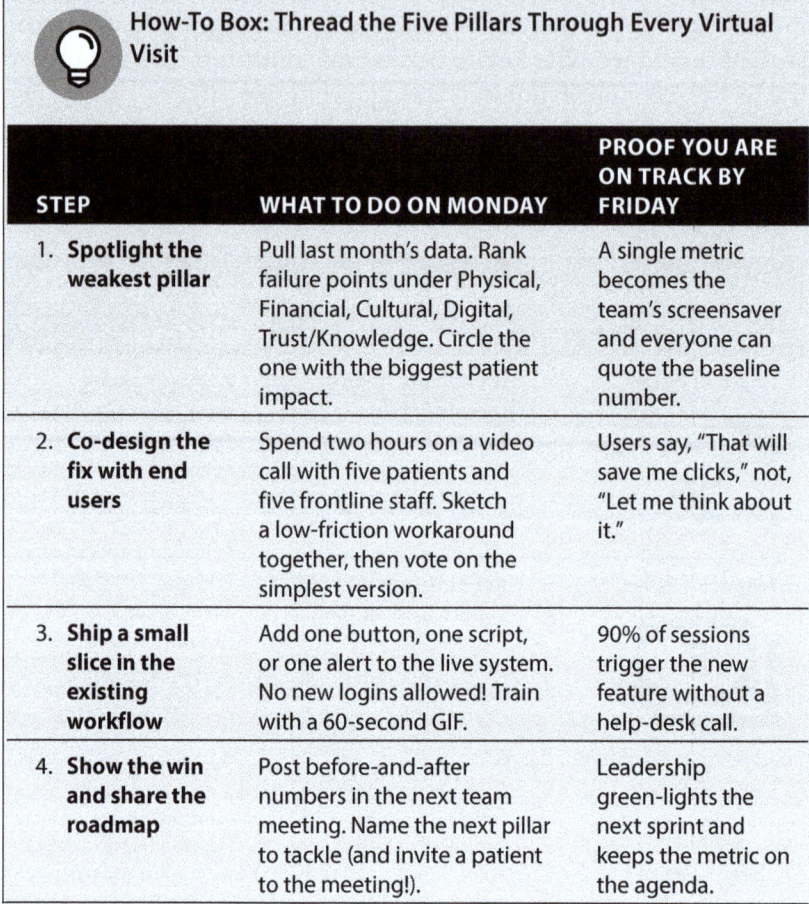

How-To Box: Thread the Five Pillars Through Every Virtual Visit

STEP	WHAT TO DO ON MONDAY	PROOF YOU ARE ON TRACK BY FRIDAY
1. **Spotlight the weakest pillar**	Pull last month's data. Rank failure points under Physical, Financial, Cultural, Digital, Trust/Knowledge. Circle the one with the biggest patient impact.	A single metric becomes the team's screensaver and everyone can quote the baseline number.
2. **Co-design the fix with end users**	Spend two hours on a video call with five patients and five frontline staff. Sketch a low-friction workaround together, then vote on the simplest version.	Users say, "That will save me clicks," not, "Let me think about it."
3. **Ship a small slice in the existing workflow**	Add one button, one script, or one alert to the live system. No new logins allowed! Train with a 60-second GIF.	90% of sessions trigger the new feature without a help-desk call.
4. **Show the win and share the roadmap**	Post before-and-after numbers in the next team meeting. Name the next pillar to tackle (and invite a patient to the meeting!).	Leadership green-lights the next sprint and keeps the metric on the agenda.

Telemedicine and Policy

Ultimately, the promise of virtual care extends beyond digital interactions. It offers an innovative, patient-centered vision of healthcare. For virtual care to achieve widespread effectiveness, supportive policies and adaptable regulations must evolve alongside technological advancements. This holistic alignment of technology, policy, and healthcare delivery will enable virtual care to genuinely fulfill its transformative potential. In doing so, it can fundamentally reshape how healthcare is accessed and delivered globally.

Historically, regulations around licensing, reimbursement, and liability have slowed the adoption of telehealth. In the United States, many state-based medical boards still require physicians to hold a license in every state where they provide virtual care, creating administrative barriers that disproportionately affect rural and underserved populations. Similarly, reimbursement models have struggled to keep pace with digital transformation. Prior to COVID-19, U.S. Medicare and Medicaid offered limited coverage for telehealth, treating it as a secondary form of care rather than a legitimate alternative to in-person visits. Think back to the "Pilotitis" phenomenon in Uganda (*See Chapter 3*) and their memorandum on pilots. This illustrates the importance of cohesive policy in telehealth implementation. It also shows what can go wrong when there is no overarching strategy for an entire system.

The future of telehealth policy must proactively tackle emerging challenges to ensure virtual care remains accessible, effective, and balanced. One critical issue that policymakers face is developing cross-border licensing models. These frameworks would allow medical specialists to consult across state and national boundaries, significantly expanding access to specialized care for underserved and remote populations globally.

Equally important is the necessity to update liability protections tailored explicitly for digital care. Virtual consultations carry unique risks not present in traditional in-person visits, and existing medical liability standards may not fully address these emerging scenarios. Clear, comprehensive liability protections can help providers confidently adopt telehealth without unnecessary legal uncertainty.

For telehealth to sustainably integrate into mainstream healthcare delivery, policymakers must also maintain reimbursement policies beyond the temporary waivers implemented during the COVID-19 pandemic. Sustainable financial support **(Financial Pillar)** ensures that virtual healthcare services remain economically viable, encouraging continuous investment and innovation in new solutions.

Ultimately, the success of virtual care depends on addressing all Five Pillars of Healthcare Access. It requires a cohesive system-wide alignment across policy, reimbursement structures, and digital infrastructure. Policymakers play a pivotal role in shaping the landscape of virtual care, and their decisions impact the extent and effectiveness of telehealth adoption globally. Several international telehealth models have already demonstrated the potential of regulatory clarity and robust financial backing to significantly drive telehealth utilization and improve healthcare accessibility worldwide (see Diagram 4.7).

REGION	REIMBURSEMENT/ LICENSING STRATEGY	IMPACT
France[10]	Telehealth visits reimbursed by Statutory Health Insurance at the same rate as in-person care. Self-employed doctors are allowed to conduct up to 20% of their activity through teleconsultations.	Ensures that telehealth services are compensated equally to in-person visits, promoting inclusive access to healthcare services across the country.
Australia[11]	National telehealth frameworks reimburse virtual visits equal to in-person visits by ongoing Medicare Benefits Schedule (MBS) arrangements.	Promotes universal healthcare access nationwide and increased adoption of telehealth services.
Germany[12]	Physicians can prescribe reimbursable digital health apps. New apps can be provisionally reimbursable by the statutory health insurance. Manufacturers must prove their app improves patients' healthcare and can negotiate long-term reimbursement.	Facilitates speed of integration of telehealth and digital solutions into regular patient care, increasing adoption.
South Korea[13]	Have allowed telehealth visits on an interim basis, as they look toward long-term regulatory changes.	Increases availability of care to an aging population.

[10] OECD. *France Country Health Profile 2023*. Published December 15, 2023. Accessed March 10, 2025. https://www.oecd.org/content/dam/oecd/en/publications/reports/2023/12/france-country-health-profile.pdf.

[11] Australian Digital Health Agency. *Telehealth*. Accessed March 10, 2025. https://www.digitalhealth.gov.au/healthcare-providers/initiatives-and-programs/telehealth.

[12] Federal Ministry of Health (Germany). *Digital Healthcare Act (DVG)*. Accessed March 10, 2025. https://www.bundesgesundheitsministerium.de/en/digital-healthcare-act.html.

[13] International Trade Administration. *South Korea – Telehealth*. Published August 2, 2022. Accessed March 10, 2025. https://www.trade.gov/market-intelligence/south-korea-telehealth.

REGION	REIMBURSEMENT/ LICENSING STRATEGY	IMPACT
Brazil[14]	Expanded telehealth frameworks include mental health, nursing, and primary care.	Expands primary and mental health services as well as care inside and outside of hospitals.
Several U.S. States[15]	38 states (and D.C.) offer some type of exception to licensing requirements for telehealth. 22 states (and the Virgin Isl.) offer special registration or licensure processes as an alternative.	Enhances telehealth adoption and ability of providers to care for patients across state lines, while continuing to have oversight.

Diagram 4-7: Global Telehealth Reimbursement and Licensing Strategies.

Each of these examples reinforces a common theme: when policy and technology align, telehealth adoption accelerates. At the same time, challenges persist. All over the world, regulatory and licensure issues have limited access in one way or another. But we have seen a huge acceleration in their evolution due to the COVID-19 pandemic. In the United States, the expansion of Medicaid-covered telehealth visits during COVID-19 temporarily improved access, but ongoing uncertainty about permanent reimbursement threatens long-term stability. Similarly, licensing restrictions prevent many specialists from providing care beyond their home state, limiting access for patients in provider-scarce regions. For innovation to flourish, policymakers must recognize telehealth as a core component of modern healthcare, not a temporary fix.

Policy at the system level determines reimbursement and the ability for providers to practice across state lines. There are also implications at the patient level.

---

[14] DLA Piper. *Telehealth Around the World: Brazil—Regulation of Telehealth*. Accessed March 10, 2025. https://www.dlapiperintelligence.com/ telehealth/countries/index.html?t=02-regulation-of- telehealth&c=BR.

[15] Center for Connected Health Policy. *State Telehealth Laws and Reimbursement Policies: Executive Summary – Fall 2024*. Published November 2024. Accessed March 10, 2025. https://www.cchpca.org/2024/11/Fall2024_Executive SummaryFINAL.pdf.

Stories from the Field: Cultural Barrier Spotlight[16]

Danny Gladden, an experienced mental health professional, social worker, and leader of Oracle Health's Behavioral Health business line, shared a vivid story about healthcare access challenges in rural Alaska. His experience highlights the complexities of integrating digital health solutions into underserved communities.

Danny served remote island communities in Alaska, where the practicalities of accessing mental healthcare were surprisingly complicated. Despite having the technology to conduct telehealth visits, regulations from the Indian Health Service required covered indigenous patients to physically travel to designated healthcare facilities to connect virtually with providers. This was the case even if those providers were comfortably sitting at home. For Danny, this meant patients from his three-island coverage area had to take a floatplane or boat to a central location just to access telehealth.

One stark memory stood out for him: The Medicaid-funded floatplanes ferrying patients who had no alternative transport to mental health appointments. This wasn't just inconvenient, it cost millions and disrupted lives. Danny described vividly how residents would lose an entire day, affecting their work, families, and daily responsibilities, simply to attend a short telehealth visit (and the doctor wasn't even there!). The irony wasn't lost on Danny. "Imagine telling someone they have to take a boat and plane to log on to a video call," he noted dryly, highlighting a stark contrast between digital potential and on-the-ground reality. "It was telehealth, but it certainly wasn't virtual care."

> **Imagine telling someone they must take a boat and plane to log on to a video call.**

Yet, the **Cultural** and **Trust/Knowledge Barriers** were even more pronounced. Some residents appreciated leaving their islands for privacy reasons—traveling provided confidentiality—but others saw it as an insurmountable burden. The indigenous community had a strong stigma surrounding mental health services, intensifying their hesitancy to seek care, especially when it meant a highly visible and burdensome trip. Danny's story underscores the essential point: effective telehealth demands more than technology. It requires thoughtful policy that removes needless barriers, culturally informed strategies that recognize and respect community sensitivities, and practical

---

[16] Gladden D. Interview with Sarah Matt. March 7, 2025.

solutions that account for the lived realities of patients. Telehealth, Danny's experience reminds us, must be more than just remote; it must genuinely meet patients where they are.

Innovation in Action: Real-World Digital Health Implementations

The Refugee Health Crisis and Virtual Care Solutions[17]

While the cold winters of Alaska can be difficult for natives, refugees in Syracuse, NY from African counties have never seen a true winter before. What's worse than the weather in Syracuse is getting your Medicaid figured out! The first experience of the U.S. healthcare system for most newly arrived refugees isn't a doctor's office or a hospital; it's a piece of mail. A Medicaid card, printed in English, arriving at a place they barely recognize as home. The **Physical Pillar** has not yet become a factor, because before they can even attempt to see a doctor, they must first navigate the labyrinth of insurance and bureaucracy.

Dr. Andrea Shaw, who oversees the Upstate Medical University Center for International Health in Syracuse, NY sees this moment of confusion unfold every day. "In the refugee camp, there was no concept of health insurance," she explains. "They either paid out of pocket or didn't get care at all. Then suddenly, they arrive in the United States, and they have to figure out Medicaid, managed care plans, provider networks, and pharmacies. Many don't even understand what a 'prescription' is."

For many refugees, their medical conditions—untreated chronic diseases, infections, or injuries from violence and displacement—demand immediate intervention. But the obstacles begin before they ever reach a doctor. The **Financial Barrier** is a major challenge, even with Medicaid, because recertification and eligibility shifts can leave them without coverage at unpredictable times. Dr. Shaw's clinic prioritizes those with urgent medical needs, but even in a state like New York, where Medicaid is more expansive than most, refugees still struggle. "We get people who settled in places like North Carolina, but once they lost Medicaid after nine months, they had no options for their child with cerebral palsy," Dr. Shaw says. "So, they pack up and move here because they hear New York has expanded Medicaid."

---

[17] Shaw A. Interview with Sarah Matt. Healthcare Access and Telehealth for Refugee Populations. February 2025.

The paradox of refugee healthcare is stark: they qualify for medical insurance, yet many still cannot access care because of **Digital Barriers**. Medicaid enrollment is done online, and notifications arrive via mail or digital portals, often in English. Even when assigned a provider, they may not be able to schedule an appointment, understand their benefits, or find a specialist who takes their insurance.

So, can telehealth help? Virtual care has the potential to bridge the gap between refugees and an unfamiliar healthcare system. Telehealth interpreters, mobile apps with multilingual interfaces, and digital scheduling platforms are beginning to transform refugee health-care. In theory, improving **Digital Barriers** should improve care by also removing logistical barriers like transportation and long wait times. The Center for International Health uses telehealth extensively, connecting specialists with patients who might otherwise never be able to see one. Mobile health apps provide medication reminders in native languages, helping patients navigate prescription regimens that were previously incomprehensible.

However, for digital health to succeed, the infrastructure must be in place. Many refugees do not have reliable Internet, smart-phones, or a private space where they can comfortably engage in virtual visits. A telehealth appointment does not work if the patient cannot connect due to poor Wi-Fi or if they do not own a compatible device.

Even when those technical issues are solved, **Cultural Barriers** present another barrier. Many refugees are unfamiliar with virtual medical visits and may feel that speaking with a doctor through a screen is ineffective or impersonal. Some come from countries where the medical hierarchy dictates that an in-person examination is the only way to receive legitimate care. If they are unfamiliar with tele-health, they may be skeptical or refuse to use it altogether.

Beyond technology, the biggest challenge remains the **Trust/ Knowledge Barrier**. The U.S. healthcare system is complex for native-born citizens, let alone for those who arrive with no concept of private insurance, provider networks, or medical bureaucracy. The **Trust/Knowledge Barrier** factor is particularly critical: if a patient doesn't trust the system, they will not use telehealth, even if it is available.

Medicaid coverage must be renewed annually, but many refugees lose it unexpectedly. "We've seen patients who arrived six months ago and then suddenly lose Medicaid without any warning," Dr. Shaw says. "For refugees, the paperwork alone can be an impossible challenge. It's in English, it requires documentation they may not have, and if they don't submit it in time, their insurance is gone."

This is where telehealth has both promise and risk. If digital health tools are used effectively, they can help patients keep track of renewal deadlines, navigate recertification, and even connect with case managers who can assist with forms. But if digital health is implemented without consideration of **Trust/Knowledge Barriers**, it can further alienate patients who already feel disconnected from the system.

Refugee Telehealth Through the Lens of the Five Pillars

For digital health solutions to succeed, they must be embedded in a system that supports all Five Pillars of Healthcare Access. Right now, that system is faltering.

- **Physical Barrier:** Telehealth can eliminate geographic barriers, but only if patients have Internet access, a quiet space, and a device capable of running a virtual visit. Otherwise, virtual care simply replicates the inaccessibility of in-person visits.

- **Financial Barrier:** While Medicaid covers telehealth, its instability means that refugees are constantly at risk of losing coverage. Telehealth must be integrated into long-term, sustainable payment models to ensure continuous care.

- **Cultural Barrier:** Virtual care must be designed with cultural competency in mind. If refugees do not understand telehealth or see it as a lower-quality form of care, adoption rates will remain low. Community education and trusted local healthcare workers are essential for bridging this gap.

- **Digital Barrier:** The assumption that all patients can navigate an online system is flawed. Many refugees lack email addresses, smartphones, or the literacy needed to use patient portals. Digital health must be paired with in-person support systems.

- **Trust/Knowledge Barrier:** A lack of trust in healthcare is one of the most significant barriers to refugee engagement with telehealth. If digital care is presented as a cold, impersonal solution, it will fail. Healthcare providers must build relationships first and use digital health as a tool rather than a replacement for human-centered care.

Stories from the Field: Trust/Knowledge Pillar Spotlight The Future of Refugee Healthcare

The intersection of refugee health and digital medicine is at a critical juncture. Technology holds immense promise, but without systemic support, it risks becoming another barrier rather than a solution. The future of virtual care for refugee populations depends on integrating digital tools with robust policy reform—ensuring that Medicaid enrollment is accessible, telehealth services are reimbursed, and healthcare providers are trained to use digital platforms effectively.

Dr. Shaw's team in Syracuse has worked to fill the gaps by creating a Refugee Health Hub, a collaborative effort between healthcare providers, public health officials, and community organizations. This ensures that patients aren't forever lost in the shuffle of insurance bureaucracy. The program has streamlined care coordination and connected refugees to services more effectively. But it remains a fragile solution, reliant on continued funding and policy stability.

Healthcare is about trust. Technology can help, but if patients don't trust the system, they won't use it.

"Healthcare is about trust," Dr. Shaw emphasizes. "Technology can help, but if patients don't trust the system, they won't use it. We need a model that doesn't just hand them a Medicaid card and hope they figure it out. We need one that walks with them, ensuring they understand how to access care, use telehealth, and navigate the system." For now, her clinic continues to provide that bridge, one patient, one Medicaid renewal, and one digital connection at a time.

Fact Check

1. Globally, more than 1 billion people are immigrants, refugees, and migrants, which is one in seven of the global population.[18]

2. The Electronic Disease Notification (EDN) system is a centralized electronic reporting system that notifies U.S. health departments

---

[18] Centers for Disease Control and Prevention. *About Immigrant and Refugee Health.* Published January 31, 2025. Accessed March 6, 2025. https://www.cdc.gov/immigrant-refugee-health/about/index.html.

and screening clinics of the arrival of refugees and immigrants with health conditions requiring follow-up.[19]

3. For refugees who do have access to mental health coverage through their insurance, access to *linguistically responsive* mental health services is limited and varies significantly by state.[20]

Stories from the Field: Digital Pillar Spotlight Telehealth and the Urban Elite, a Different Kind of Access Problem[21]

While the stories of refugees and rural Alaska seem plausible to readers when we think of access barriers, most people don't even think of wealthy patients also having access barriers (see Diagram 4.8). Dr. Kedar Sankholkar, a cardiologist practicing in the heart of Manhattan, sees an interesting and diverse patient population. His office is located on 57th and 7th, what many call "Billionaire's Row." Yet, as he reflects on years of practice, he sees a striking irony: even in one of the wealthiest pockets of the world, access to healthcare remains a profound challenge.

During the COVID-19 pandemic, Dr. Sankholkar's practice pivoted quickly to telemedicine. What started as a necessity soon became a core part of his workflow. "Five years later, I use telehealth daily. Out of 15 patients, about 30 percent are virtual. For follow-ups and time-sensitive consultations, it's a game changer," he explains. But what has surprised him the most isn't telehealth's impact on low-income patients, it's how much it benefits his wealthiest ones.

> **Even in one of the wealthiest pockets of the world, access to healthcare remains a profound challenge.**

His high-net-worth patients—CEOs, financiers, and jetsetters—can afford the best medical care, yet they often fail to engage with it. The **Financial Barrier** isn't the issue here; it's time.

---

[19] Centers for Disease Control and Prevention. *Electronic Disease Notification System.* Published May 15, 2024. Accessed March 6, 2025. https://www.cdc.gov/immigrant-refugee-health/php/case-reporting-edn/index.html.

[20] U.S. Committee for Refugees and Immigrants. *Mental Health Awareness Month: Barriers and Access to Mental Health Care.* Published May 9, 2024. Accessed March 6, 2025. https://refugees.org/mental-health-awareness-month-barriers-and-access-to-mental-health-care/.

[21] Sankholkar K. Interview with Sarah Matt, MD. February 14, 2025.

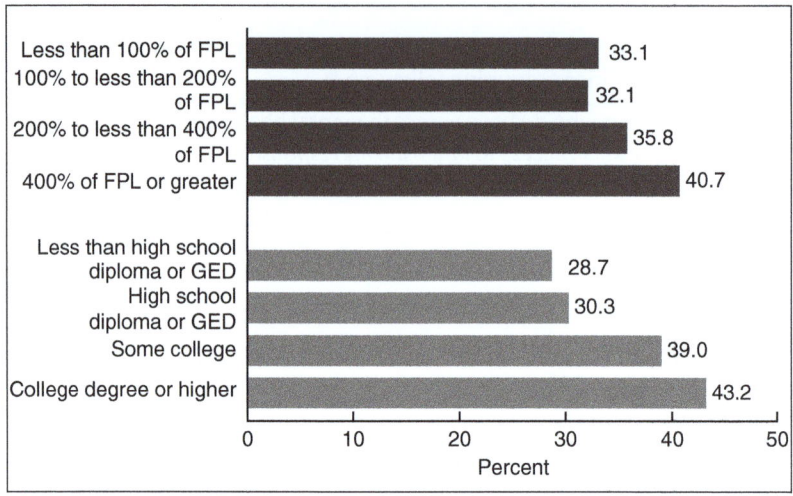

Diagram 4-8: Percentage of Adults Aged 18 and Over Who Used Telemedicine in the Past 12 Months, by Family Income and Education Level: United States, 2021 (FPL, Federal Poverty Line).[22]

"I can't tell you how many times my office has called patients 12 times to remind them about a follow-up test. No response. And then, two years later, I get a frantic call from an ER, turns out they're having an aortic dissection," he says with a mix of frustration and inevitability.

His patients aren't avoiding care because they lack money or insurance; they avoid it because they are too busy, too distracted, or too focused on everything except their own health. **Cultural Barriers** play a role, as many believe that if they feel fine, they don't need medical attention. Dr. Sankholkar sees telehealth as a bridge between urgency and inaction. "It allows me to meet patients where they are, whether it's in a corporate office between meetings or in a shelter, where getting to an appointment is logistically impossible. The more barriers we remove, the better outcomes we'll see." His billionaire patients don't need financial help; they need care to be delivered in a way that fits into their lives. Yet, he is quick to add a caveat. "Tech isn't a magic fix. It's just a tool. It only works if we design systems that fit into people's lives, not the other way around." The **Digital Pillar** is not just about having the technology, it's about using it in a way that aligns with behavior, trust **(Trust/Knowledge Pillar)**, and urgency.

[22] Lucas JW, Villarroel MA. *Telemedicine Use Among Adults: United States, 2021.* NCHS Data Brief, no. 445. Hyattsville, MD: National Center for Health Statistics. 2022. doi:10.15620/cdc:121435.

Stories from the Field: Trust/Knowledge Barrier Spotlight
A Tale of Two Cities (Same Barriers, Different Reasons)[21]

Whether they're struggling to pay rent or running a Fortune 500 company, people deprioritize their health until it becomes a crisis.

Reflecting on his training at Beth Israel in Manhattan, Dr. Sankholkar recalls working with Medicaid patients and individuals experiencing homelessness. Despite stark differences in wealth, he observed a common thread: "Whether they're struggling to pay rent or running a Fortune 500 company, people deprioritize their health until it becomes a crisis." In both cases, **Trust/ Knowledge Barriers** serve as major obstacles. For the wealthy, the belief that money alone insulates them from health risks leads to dangerous delays in care. For low-income patients, distrust in the healthcare system, fear of unexpected costs, and logistical hurdles prevent them from engaging with care early.

And then, of course, there are **Physical Barriers**. One patient might be unable to take off work to travel across the city for an appointment, while another might physically struggle to get to a clinic. Telehealth alleviates some of these barriers, but only if patients use it.

The Telehealth Paradox

Dr. Sankholkar's experience highlights a crucial insight: Telehealth is not just for those in rural areas or underserved communities; it is equally vital for urban populations, including the ultra-wealthy. Yet the reasons for needing telehealth differ dramatically:

- For the low-income patient, it removes logistical, financial, and transportation barriers.
- For the high-net-worth patient, it removes the barriers of time, distraction, and perceived invincibility.

Regardless of income level, access to care is about more than just **Financial Barrier** resources, it's about engagement, trust **(Trust/ Knowledge Barrier)**, and how seamlessly healthcare fits into someone's daily life. Telehealth, when designed thoughtfully, is not just a service; it is a means of making care accessible to everyone, whether they're dialing in from a high-rise office or a crowded shelter.

Next Shift Quick Wins

- Text patients at T-30 minutes with a one-tap "I am running late" button.
- Auto-offer the next open virtual slot and reschedule with no staff clicks.
- Push a same-day fill-in alert to the waitlist. Keep slots at 95 percent utilization.

Pushing the Boundaries of What's Possible

Historically, telehealth has largely been confined to basic video visits; a useful but limited substitute for in-person care. But let's remember that many times a virtual visit is not replacing an in-person visit, it's actually replacing no care at all. Today the horizon of virtual healthcare is expanding far beyond mere video interactions, evolving toward a future characterized by intelligent, data-driven, proactive healthcare that fundamentally transforms patient management and outcomes.

Artificial intelligence (AI) now plays a pivotal role in reshaping remote healthcare. AI-driven diagnostic technologies are not only complementing but also enhancing the accuracy and efficiency of medical assessments. Advanced algorithms analyze complex imaging scans, revealing subtle indicators of early-stage diseases like cancer, cardiovascular conditions, or neurological disorders that even seasoned clinicians might miss. AI-supported *remote patient monitoring* (RPM) has revolutionized routine healthcare devices, such as blood pressure cuffs, glucose monitors, and pulse oximeters, by converting them into continuous streams of action-able health data. These devices proactively alert healthcare providers to critical changes in a patient's condition, enabling timely interventions before conditions worsen. *(I dig more deeply into AI in Chapter 6.)*

Telesurgery, previously considered a distant possibility, is rapidly becoming a reality as well. Companies like Sovato are bridging the gap by empowering robotic surgical systems to operate remotely. This technological leap is poised to break down geographical **Physical Barriers**, ensuring patients worldwide have access to specialized surgical expertise without the limitations imposed by physical location.

Yet, these innovations prompt significant ethical and regulatory considerations. Key questions arise around decision-making authority.

Should an AI's alert be the decisive factor in clinical interventions, or should it strictly be an informational tool supporting clinician judgment? If an AI says that I have high blood pressure, do I argue with the algorithm? "Hey Siri, how do you feel about lowering my salt intake?" Equally important is the issue of bias within AI algorithms. Incomplete or non-diverse datasets may inadvertently reinforce disparities in healthcare outcomes. Protecting patient privacy amidst increasingly interconnected health platforms further complicates this landscape, demanding a careful balance between innovation and confidentiality.

Hypothetical Scenario: AI-Powered Remote Monitoring in Action

Despite potential risk, there are also infinite possibilities when it comes to AI remote patient monitoring (RPM). Imagine a patient living with hypertension and early-stage heart failure. They diligently take medication yet experience fluctuating blood pressure levels. Through an AI-enhanced RPM program, the patient utilizes a smart blood pressure cuff and wearable monitoring technology, continually capturing vital signs and transmitting this data to an intelligent analytic system.

One morning, despite the patient feeling "normal," the system identifies a subtle but significant pattern, an elevation in blood pressure accompanied by irregular heart rhythms. The AI immediately alerts the patient's healthcare provider. Within hours, a nurse initiates contact, followed by a same-day virtual consultation with their physician, who promptly adjusts medication dosage. This rapid, proactive intervention effectively prevents what could have otherwise led to an emergency hospitalization.

Such scenarios demonstrate the transformative potential of AI-driven virtual care, not merely reacting to health crises but actively anticipating them. As early as 2014, an Australian telemedicine study of chronic kidney disease patients demonstrated remarkable cost reductions for patients utilizing RPM. These savings consisted of 46 percent in Medicare Benefits Schedule expenditures and 25 percent in Pharmaceutical Benefits Scheme expenditures.[23] Now, more

---

[23] Australian Digital Health Agency. *Australia's National Digital Health Strategy: Safe, Seamless and Secure*. Published August 2017. Accessed March 9, 2025. https://
www.digitalhealth.gov.au/sites/default/files/2020-11/
Australia%27s%20National%20Digital%20Health%20Strategy
%20-%20Safe%2C%20seamless%20and%20secure.pdf.

than a decade later, integrating advanced AI technology into virtual healthcare has magnified this potential, promising even greater efficiencies and outcomes.

Recent clinical trials further underscore this promise. A study of over 900 adults discharged after non-elective surgery in Canada demonstrated that *remote automated monitoring* (RAM), paired with virtual nursing interactions, significantly improved clinical outcomes.[24] Although RAM did not drastically change the number of days patients spent alive at home, it substantially enhanced the detection and correction of medication errors and significantly reduced patient-reported pain levels.[25] As AI and RPM/RAM continue to evolve, their impact on healthcare will undoubtedly intensify. This ongoing innovation is not merely reshaping virtual healthcare; it is redefining what is possible. This is paving the way for a future where healthcare is accessible, proactive, and tailored precisely to patient needs.

Fact Check

1. The telemedicine market is predicted to grow at a CAGR (compound annual growth rate) of 24.02 percent during 2020–2023.[26]

2. Specialists treating chronic conditions are the main users of RPM. Internal medicine (28.7%), cardiologists (21.3%), and family

---

[24] McGillion MH, Parlow J, Borges FK, et al. Post-discharge after surgery virtual care with remote automated monitoring-1 (PVC-RAM-1) technology versus standard care: randomized controlled trial. *BMJ.* 2021;374:n2209. Accessed March 10, 2025. https://www.bmj.com/content/374/bmj.n2209.

[25] McGillion MH, Parlow J, Borges FK, et al. Post-discharge after surgery virtual care with remote automated monitoring-1 (PVC-RAM-1) technology versus standard care: randomized controlled trial. *BMJ.* 2021;374:n2209. Accessed March 10, 2025. https://www.bmj.com/content/374/bmj.n2209.

[26] Telehealth and telemedicine research 2020–2024, 2025–2030 with analyst recommendations – Integration of wearable devices and the IoT develop strategies for expanding adoption in emerging markets – ResearchAndMarkets.com. *Business Wire.* Published January 10, 2025. Accessed March 9, 2025. https://www.businesswire.com/news/home/20250110337200/en/Telehealth-and-Telemedicine-Research-2020-2024-2025-2030-with-Analyst-Recommendations---Integration-of-Wearable-Devices-and-the-IoT-Develop-Strategies-for-Expanding-Adoption-in-Emerging-Markets---ResearchAndMarkets.com.

practitioners (19.4%) account for the most claims in the United States.[27]

3. By 2030, experts predict that chronic diseases will contribute to as much as 84 percent of total global mortality.[28]

 Next Shift Quick Wins: Portal Activations Before the Weekend

▪ Print wallet-size QR cards at check-in. Have staff scan and activate accounts on the spot.

▪ Offer a $5 cafeteria credit for first-time logins. Track redemption to measure uptake.

▪ Post a leaderboard at the nursing station. Celebrate the unit that tops weekly activations. (Tacos are the perfect reward!)

Virtual Care as the New Standard

Picture a world where distance no longer dictates the quality of health-care you receive. Where a patient in a remote farming community gets the same specialist care as someone in a bustling city, where a person's ZIP code no longer determines their health outcomes. This is the promise of virtual care—a promise that is no longer just theoretical, but one we are actively building toward. For virtual care to truly become the standard, the five pillars must be solidly in place:

Physical Pillar: Virtual care cannot be an urban luxury. It must stretch beyond city centers to rural areas, underserved communities, and places where medical deserts have long existed. If broadband stops at the edge of town, so does access to care.

---

[27] TATEEDA GLOBAL. *The Future of Telehealth: Trends and Innovations in Telemedicine in 2025 and Beyond.* Accessed March 10, 2025. https://tateeda.com/blog/future-of-telehealth.

[28] McKinsey & Company. *The Health Benefits and Business Potential of Digital Therapeutics.* Accessed March 10, 2025. https://www.mckinsey.com/industries/life-sciences/our-insights/the-health-benefits-and-business-potential-of-digital-therapeutics.

Financial Pillar: If the reimbursement structure for virtual care remains unstable or temporary, its adoption will be slow and unsustainable. Healthcare systems and policymakers must commit to funding models that support telemedicine as a legitimate, reimbursable form of care, not just an emergency backup.

Cultural Pillar: A telehealth visit is not just about technology; it is about trust. For virtual care to be effective, it must meet patients where they are—linguistically, socially, and culturally. A one-size-fits-all approach will not work in a world as diverse as ours.

Digital Pillar: A video visit is only useful if the patient and provider can connect. This means investing in infrastructure, broadband expansion, reliable technology, and platforms that work for patients of all digital literacy levels. Otherwise, virtual care risks widening disparities rather than closing them.

Trust/Knowledge Pillar: The best virtual care program in the world is meaningless if patients and providers do not believe in it. Education is key, not just for patients navigating telehealth for the first time, but for clinicians, hospital administrators, and policymakers who must integrate virtual care into everyday practice.

AI and the Future of Virtual Care

As telehealth transitions from convenience to necessity, AI sits at the heart of this transformation. The future of virtual healthcare transcends simple remote consultations. It is moving toward predictive, personalized, and continuous care models that maintain wellness and prevent hospitalizations. Yet, to ensure these innovations genuinely benefit all patients, we must tackle significant challenges head-on. Inclusive access, stringent ethical safeguards, and trust in AI-driven decision-making are essential for realizing this potential responsibly.

Virtual encounters will never replace the reassuring warmth of a physician's hand. Yet, when built on fair access and anchored in sound policy, they can bring that same expertise into every living room, barber shop, and synagogue community room. As we move to robotics and artificial intelligence in the next chapters, remember that no algorithm will succeed unless it first travels the same five pillars. Real progress will belong to teams that join silicon precision with human empathy, one pixel-perfect visit at a time.

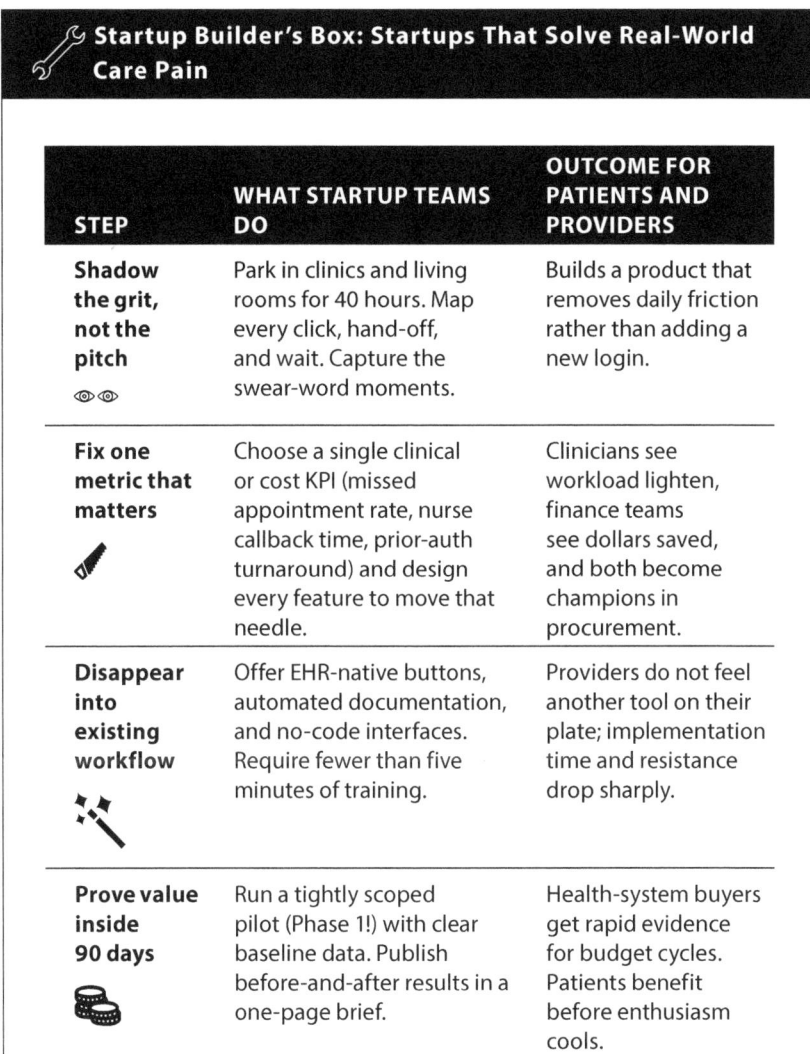

Startup Builder's Box: Startups That Solve Real-World Care Pain

STEP	WHAT STARTUP TEAMS DO	OUTCOME FOR PATIENTS AND PROVIDERS
Shadow the grit, not the pitch	Park in clinics and living rooms for 40 hours. Map every click, hand-off, and wait. Capture the swear-word moments.	Builds a product that removes daily friction rather than adding a new login.
Fix one metric that matters	Choose a single clinical or cost KPI (missed appointment rate, nurse callback time, prior-auth turnaround) and design every feature to move that needle.	Clinicians see workload lighten, finance teams see dollars saved, and both become champions in procurement.
Disappear into existing workflow	Offer EHR-native buttons, automated documentation, and no-code interfaces. Require fewer than five minutes of training.	Providers do not feel another tool on their plate; implementation time and resistance drop sharply.
Prove value inside 90 days	Run a tightly scoped pilot (Phase 1!) with clear baseline data. Publish before-and-after results in a one-page brief.	Health-system buyers get rapid evidence for budget cycles. Patients benefit before enthusiasm cools.

Leaders' End-of-Chapter Action Checklist: Chapter 4: "Virtual Care Unleashed"

LEADER	HIGH-IMPACT ACTION THAT MOVES VIRTUAL CARE FROM PILOT TO CORE
❑ Board Director	Approve a public Virtual Care Scorecard. Tie 5% of the annual capital plan to progress on uptime, first contact resolution, and patient show rate
❑ Chief Executive Officer	Commit to a Virtual First policy that offers 90% of primary-care appointments within 24 hours; review progress in monthly operating meetings
❑ Chief Information Officer	Deliver 99.9% video-visit uptime; display a live traffic-light dashboard for clinicians and escalate outages within five minutes
❑ Chief Health Information Officer	Embed evidence-based templates that auto-populate billing codes and patient instructions; target a 30% reduction in clinician documentation time
❑ VP Clinical Operations	Add an AI-driven triage layer that diverts routine visits to video; aim to cut avoidable emergency transfers by 15% in six months
❑ VP Nursing and Patient Education	Launch Virtual Welcome Calls for new patients; resolve 75% of tech questions before the first visit
❑ VP Data and Analytics	Monitor patient-flow algorithms for drift and privacy risks; send quarterly reports with mitigation playbooks to the Quality Committee
❑ Telehealth Program Manager	Release monthly micro-learning clips on camera etiquette and sound checks; sample 20 encounters to confirm adoption
❑ Patient Experience Manager	Install one-tap emoji feedback in the portal and on-site kiosks; circulate a Weekly Wins and Fixes reel to all frontline teams
❑ Community Health Worker Supervisor	Host digital-literacy pop-ups in libraries and laundromats; chase a 10% jump in portal activation within four months
❑ Director of Snacks and Morale	Deliver latency-free lattes and packet-loss pastries to every sprint review; include a QR code poll asking whether espresso improves error-rate detection

The Robotics Revolution

The Moment Everything Changed

I remember when the headhunter called me the first time to ask me to join the Sovato team. The team is amazing, the founders are solid, and they are enabling remote robotic surgery! Remote robotic surgery? Hilarious. Thanks for the call, bud. My first thought was, who would trust a robot holding a scalpel? Turns out, more people than I thought!

I did not have much intention of even thinking about remote robotic surgery. Truly. It seemed too far away, with too many barriers to system integration, regulatory and reimbursement hurdles, and the trust of patients and surgeons. It was going to be an uphill battle. But that guy was a trooper, and he kept calling me back again and again. He was very persistent. I met the CEO, I met the cofounder, I met the head of engineering, and the Chief Medical Officer (he was my fav!). I heard their story, I heard their passion, and I finally saw the opportunity for patients, providers, and global healthcare. And I got excited. The path was not to have "robot surgeons" (thank goodness for that!). The point was that the distance between a surgeon and the patient wasn't going to be measured in inches anymore, but in miles.

This was my first experience with remote robotic surgery, and I remember the exact moment I realized everything was about to change. The precision, the control, the seamless integration of technology, the potential for improved access, and new markets. It was unlike any opportunity I had encountered. In that instant, I knew that virtual care was not just a convenience; it was the future of medicine. And virtual care wasn't just UTIs and well visits—it could be everything, even surgery.

> Virtual care isn't just UTIs and well visits—it can be everything, even surgery.

My Robotic Journey: From Skepticism to Leading the Charge

The first groundbreaking remote robotic surgery, famously known as the "Lindbergh Operation," took place on September 7, 2001, when surgeons in New York successfully removed the gallbladder of a patient in Strasbourg, France.[1] This milestone demonstrated not only the technological feasibility of telesurgery but also its potential to redefine surgical care access globally. It showed that expert surgical care could transcend distances **(Physical Barrier)**, offering hope to underserved patients everywhere. However, the Lindbergh Operation required sophisticated technological infrastructure, including high-speed fiber-optic connections, advanced robotic systems, and real-time communication protocols, highlighting significant barriers to widespread adoption. Essentially it was the most expensive gallbladder removal ever! (Even the gallbladder left with sticker-shock!)

I didn't start as a robotics enthusiast, quite the opposite. Initially skeptical, I saw robotic surgery as expensive, overly complicated, and detached from the human touch integral to medicine. But over years of seeing technologies develop, especially during my tenure at Oracle and at several startups, skepticism gave way to excitement. Working directly with innovators and seeing how robotic systems could bridge vast distances and deliver specialized care to underserved areas converted me into an advocate.

---

[1] Marescaux J, Leroy J, Rubino F, et al. Transatlantic robot-assisted telesurgery. *Nature.* 2001;413(6854):379–380. Accessed June 22, 2025. https://www.nature.com/articles/35096636.

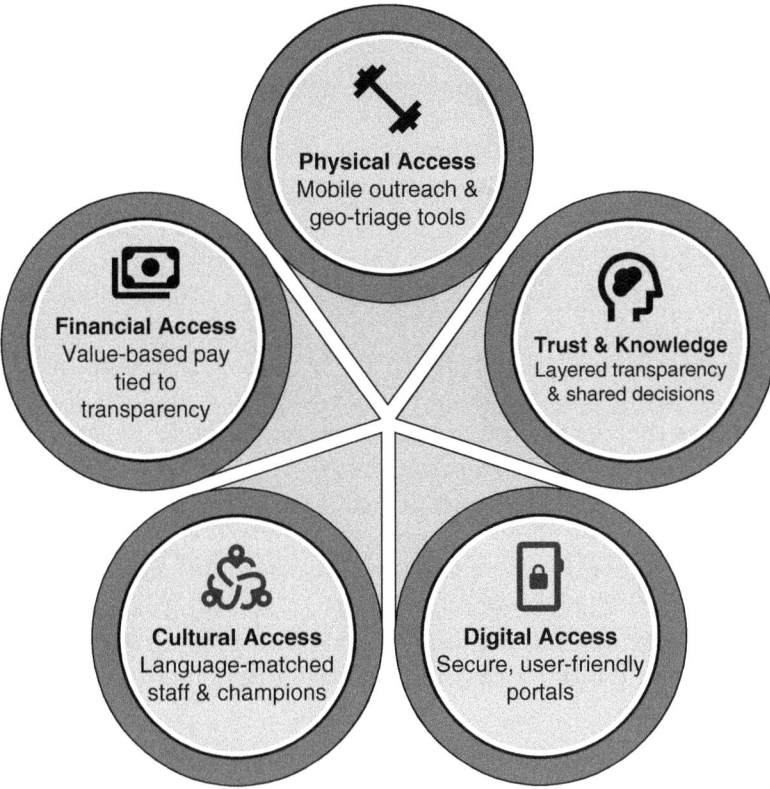

Diagram 5-1: Pillar Spotlight.

My experience at Oracle Health integrating large-scale digital initiatives had already taught me the transformative potential of technology. But robotics was different. It wasn't just transformative; it was revolutionary. While there are companies around the globe focused on building the best robotics for surgical use, and there are hospitals and ministries of health all over the world focused on care delivery, there was still a gap. Bringing the best of modern innovation could not be done alone and needed both of these groups. Glue was needed to ensure the Five Pillars of Access were addressed (see Diagram 5.1). The right infrastructure, the right sustainable funding models, the right training and education, and much more, was necessary to produce successful and safe surgical robotics programs at scale. With the advent of remote robotic surgery, these pillars are even more important, as they touch multiple locations at the same time.

Dr. Dennis Fowler, a pioneer in minimally invasive surgery, Chief Medical Officer at the startup Sovato, and a man I would call a friend, initially approached robotic surgery with skepticism. "We were against it," he shared candidly, reflecting a sentiment common among many of his peers at the advent of robotic surgery, who believed robotic solutions were costly and perhaps unnecessary.[2] Yet over time, Fowler observed an undeniable shift. Despite personally training numerous surgeons to mastery with traditional laparoscopic techniques, he found that each one eventually adopted robotic surgery as their modality of choice as their career progressed. It wasn't merely about clinical outcomes or reduced complications; the transition was driven by ergonomics, reduced physical strain, and ultimately, the intuitive control that robotics offered, providing an exceptional experience for the surgeon. Fowler's story exemplifies a broader shift: how surgeons initially resistant to change become enthusiastic adopters once they experience robotics firsthand, overcoming skepticism not by coercion but by undeniable benefit to their practice and their patients.

Historical Context and Evolution of Robotic Surgery

My own medical career, driving across rural Texas performing house calls, reminded me vividly of healthcare's origins—doctors traveling to patients, bridging **Physical Barriers**. My experiences involved modern-day inconveniences like stubborn donkeys blocking country roads, and Whataburger running out of jalapenos, rather than horse and buggy issues. (Pro tip: no amount of clinical empathy moves an angry donkey!) But the core challenge remained unchanged: overcoming geography to deliver care. While robotic surgery and the potential of remote robotic surgery to improve patient access are now in arms reach, it's worth remembering that improving healthcare access has always been an evolving challenge (see Diagram 5.2).

---

[2] Fowler D. Interview with Sarah Matt, MD. March 21, 2025.

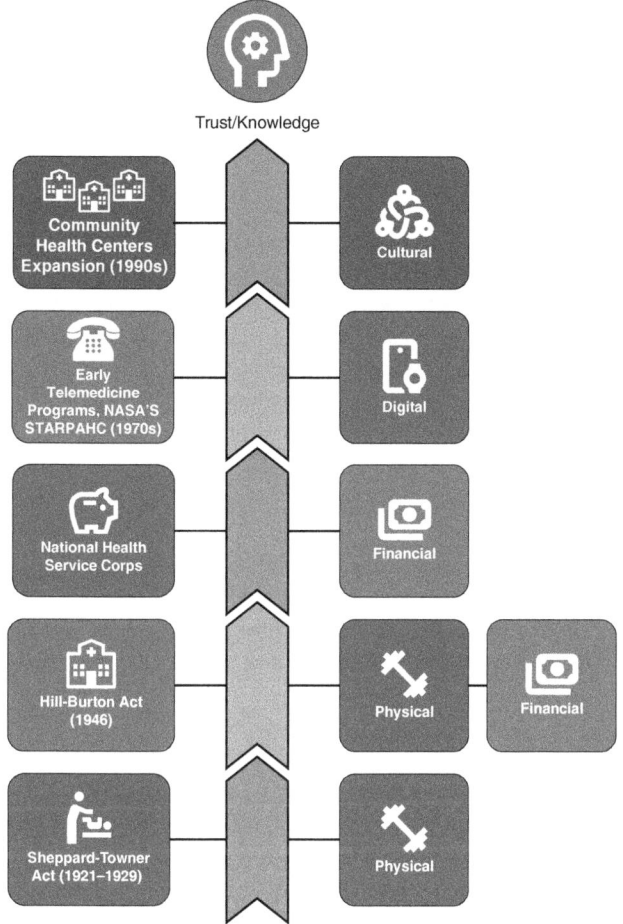

Diagram 5-2: Timeline of U.S. Healthcare Access Improvements.

Fact Check

- **Sheppard-Towner Act (1921–1929): Federal funding for prenatal and child health centers.[3]**
  - The funding appropriated was $5,000 per state in annual grants and $1.2 million in matching federal funds, with states' participation entirely voluntary.

(continues)

---

[3] Kaiser Family Foundation. *History of Health Insurance Benefits*. Published March 2011. Accessed January 20, 2025. https://www.kff.org/wp-content/uploads/2011/03/5-02-13-history-of-health-reform.pdf.

(continued)

- Public nurses made over three million visits to the homes of women with infants during the time the act was in effect.[4]
- Reduced infant mortality rates in participating states.
- **Hill-Burton Act (1946):**[5] Built thousands of hospitals across rural America.
  - A total of 10,490 projects were funded in mainland United States at a total federal cost of $27.3 billion (in 2012 dollars).
  - Counties that received funds saw large increases in nonprofit and public capacity and in those same counties, for-profit hospitals decreased capacity, converted to nonprofit status, or exited altogether. This had a lasting and profound impact on the composition of the hospital industry.
- **National Health Service Corps (1970s):**[6] Provided financial incentives for healthcare workers in underserved areas.
  - Since 1972, more than 80,000 physicians, dentists, behavioral health professionals, and women's health clinicians have completed NHSC service.
  - In 2024, the NHSC was able to award just 5 percent of eligible scholarship applications and half of eligible loan repayment applications.
- **Early Telemedicine Programs (1970s):**[7] NASA's STARPAHC project pioneered remote healthcare delivery.
  - Delivered remote healthcare to the Tohono O'odham reservation, treating patients via two-way TV links.
  - The State of Arizona was one of only a handful to allow the delivery of care by PAs/NPs, ensuring that the terrestrial experiment would more closely resemble the situation during space missions.
  - NASA has integrated telemedicine into every human spaceflight program, including the International Space Station and Artemis.
- **Community Health Centers Expansion (1990s):** Focused on social determinants of health and integrated care.

[4] Madgett K, *Sheppard-Towner Maternity and Infancy Protection Act (1921)*. Embryo Project Encyclopedia. Published May 18, 2017. ISSN: 1940-5030. https://hdl.handle.net/10776/11503.

[5] Chung AP, Gaynor M, Richards-Shubik S. *Subsidies and Structure: The Lasting Impact of the Hill-Burton Program on the Hospital Industry*. National Bureau of Economic Research. Published February 2016. Accessed January 20, 2025. https://www.nber.org/system/files/working_papers/w22037/w22037.pdf.

[6] Bureau of Health Workforce. *National Health Service Corps – 50 Years of Service.* Health Resources and Services Administration. Accessed January 20, 2025. https://nhsc.hrsa.gov.

[7] NASA. *A Brief History of NASA's Contributions to Telemedicine*. Accessed January 20, 2025. https://www.nasa.gov/wp-content/uploads/2024/03/nasatelemedicine-briefhistory.pdf.

Each of these historical efforts underscores critical lessons in creating scalable and sustainable healthcare systems. Sustainable financial models **(Financial Pillar)** have always been central; without financial stability, even the best-intentioned programs can falter. This was clearly illustrated by the Hill-Burton Act, whose investments in physical infrastructure significantly reshaped rural healthcare, and by the National Health Service Corps, which leveraged financial incentives **(Financial Pillar)** to ensure clinicians remained in underserved communities.

Yet funding alone is not sufficient. You can't have a fully functioning healthcare system just by throwing money at it. (Trust me, they have tried!) All the various components need to be executed properly. Engaging communities directly is vital for long-term success. The expansion of Community Health Centers in the 1990s demonstrated how involving local populations directly in the design and delivery of services not only improved acceptance of care but also cultivated trust. Trust **(Trust/Knowledge Pillar)** and community engagement have always gone hand in hand. Consider the Sheppard-Towner Act, where public nurses, respected and familiar faces in the community, conducted over three million home visits, providing care and education directly to mothers. These nurses became trusted community figures, whose consistent presence and guidance reduced infant mortality rates by educating mothers and ensuring healthcare aligned closely with cultural expectations.

Yet, despite such successes, workforce recruitment and retention have persistently remained challenging. Even robust programs like the National Health Service Corps in the 1970s, which provided financial incentives to attract providers to underserved areas, continue to face challenges in meeting the overwhelming demand. Infrastructure has also continually mattered, encompassing more than just physical buildings. Transportation has proven crucial time and again, as seen in efforts from early house-call doctors navigating rural landscapes to today's persistent struggles with patients who cannot afford or physically manage the travel to distant medical facilities.

Initiatives like NASA's early STARPAHC telemedicine project demonstrate that technology has always been part of the equation.[8] The historical evolution of healthcare access illustrates a recurring theme: technology and delivery models may have advanced, but the fundamental challenges of geography, funding, community engagement, and human capital remain remarkably consistent. Understanding the lessons of history provides a foundation to evaluate whether

---

[8] NASA. *A Brief History of NASA's Contributions to Telemedicine.* Accessed January 20, 2025. https://www.nasa.gov/wp-content/uploads/2024/03/nasatelemedicine-briefhistory.pdf.

today's technological breakthroughs are genuinely revolutionary or simply modern iterations built upon decades of accumulated insights.

Stories from the Field: Digital Barrier Spotlight[9]

Despite all the optimism around robotics and its potential to expand access, the reality on the ground is far more complex. I spoke recently with a surgeon working in a major metropolitan area and employed by a large, integrated health system known nationally for its structured approach to care. He had spent the earlier part of his career in a more rural setting, serving a smaller community with fewer physicians. Ironically, it was in that rural role where he had significantly more access to robotic surgical platforms than he does today.

"In my last job," he told me, "We had two systems and were expanding to a third. I could book several robotic cases a month with no problem. Here, I get one day."

The issue wasn't about surgeon readiness or patient demand; it was about hospital access, scheduling limitations, and state-level budgeting rules. His current facility, though located in a highly resourced, urban area, is not owned by his health system. Because of unique state financial models and broader competition for time across departments, robotic access is rationed, shared among all surgical disciplines.

"In a bigger city with more complex financial structures, I've had to scale back what I'm actually trained to do," he said. "It's not that the robot isn't there. It's that I can't get time on it."

This contradiction, more tech, less access, is a recurring theme. In rural areas, fewer competing surgical teams can sometimes mean greater individual access to innovation. In large urban centers, even the most advanced technologies are subject to scheduling hierarchies, contract negotiations, and legacy infrastructure decisions that limit their use.

"I moved to this system thinking it would be a step forward," he added. "And in many ways, it is. But the irony is that I'm doing fewer robotic procedures now than I was before."

His story underscores a broader truth: introducing new surgical technology doesn't automatically translate into expanded access. Physical presence of equipment is not enough. The full system—financial models, clinical operations, hospital politics, even vacation policies—must be aligned to make high-tech care universally available. Until then, even the most advanced tools risk becoming underused or unavailable in places where they're needed most.

---

[9] Anonymous surgeon employed in a large integrated health system. Interview with Matt S. April 6, 2025.

Robotic Surgery's Arc and Its New Frontier in Access

Robotic-assisted surgery began as a battlefield concept that became clinical reality in 1985 when surgeons used the PUMA 560 robot for a neurosurgical biopsy.[10] Early systems were limited. However, the 1990s transformed the field. ROBODOC won FDA clearance in 1992 in orthopedics, while Computer Motion's AESOP endoscope arm earned approval in 1994 and, two years later, it gained voice control, giving surgeons a "third hand."[11] In 2000, Intuitive's da Vinci platform secured general laparoscopic clearance, pairing three-dimensional optics with wristed instruments to restore intuitive movements and rapidly set a new standard for minimally invasive care.[12]

As the quest to expand healthcare access has evolved, so too have the technologies designed to bridge any **Physical Barriers**. Remote robotic surgery, once considered a concept from science fiction, is rapidly becoming a clinical reality. The premise is simple yet transformative; enabling a surgeon located in one geographic location to perform surgery on a patient hundreds or even thousands of miles away using advanced robotics, high-speed data connections, and immersive visual interfaces. While today's telehealth technology has vastly improved remote consultations, remote robotic surgery goes one critical step further: delivering actual physical interventions remotely.

Dr. Fowler provides a compelling paradox when discussing robotic surgery's impact on access (see Diagram 5.3).[13] He notes that while robotic procedures undeniably offer equal to superior outcomes, their availability is highly uneven. Reflecting on the example of prostate cancer care, Fowler points out stark geographic and socioeconomic disparities, particularly in rural or impoverished regions, where facilities simply can't afford robotics. He described scenarios from rural Alabama, illustrating vividly how the absence of robotics not only restricts patients from receiving optimal care but also prevents skilled

---

[10] Morrell ALG, Morrell-Junior AC, Morrell AG, et al. The history of robotic surgery and its evolution: when illusion becomes reality. *Rev Col Bras Cir.* 2021;48:e20202798. Accessed June 22, 2025. https://pubmed.ncbi.nlm.nih.gov/35445466.

[11] Intuitive. *Company History*. Intuitive website. Accessed June 22, 2025. https://www.intuitive.com/en-us/about-us/company/history.

[12] Marescaux J, Leroy J, Rubino F, et al. Transatlantic robot-assisted telesurgery. *Nature.* 2001;413(6854):379–380. Accessed June 22, 2025. https://www.nature.com/articles/35096636.

[13] Fowler D. Interview with Sarah Matt, MD. March 21, 2025.

ASPECT OF ROBOTIC SURGERY	IMPACT ON HEALTHCARE
Precision	Significantly reduces surgical errors and enhances patient safety.
Recovery	Shortens recovery times, leading to quicker discharge and fewer complications.
Clinical Possibilities	Enables complex and precise procedures previously considered challenging or impossible, thus expanding available specialized care options.
Cost Effectiveness	Despite high initial costs, reduces complications and re-admissions, contributing to overall cost savings in healthcare delivery.

Diagram 5-3: Impact of Robotic Surgery on Healthcare Access.

surgeons from practicing in underserved areas. The consequence is a troubling cycle of imbalance, where robotics becomes both a life-saving advancement for some and an unreachable luxury for others. This paradox emphasizes that while technological innovation has immense potential to bridge geographic divides, it simultaneously risks exacerbating gaps if not thoughtfully and sustainably implemented.

History shows that each breakthrough only delivers broad benefit when the other pillars—financial models, cultural acceptance, digital infrastructure, and trust—advance in parallel. Remote systems must therefore pair hardware with sustainable payment pathways, community engagement, robust training for local teams, and secure low-latency networks. When those elements mature together, robotics can finally fulfill its promise of bringing high-quality surgical care to patients rather than forcing patients to chase high-quality care.

Fact Check[14]

- The global robotic surgery market is projected to exceed $14 billion by 2026, highlighting the rapid growth and increasing reliance on automation in healthcare settings worldwide.

- Patients who undergo robotic-assisted surgery have significantly less postoperative pain, reduced open conversion rate, and

---

[14] Fay K, Patel AD. Should robot-assisted surgery tolerate or even accommodate less surgical dexterity? *AMA J Ethics*. 2023;25(8):E584–590. doi:10.1001/amajethics.2023.584.

> shorter postoperative length of stay than patients who undergo laparoscopic surgery.
>
> ■ A 2010 study found that, on average across a variety of surgical procedures, utilization of the robotic platform added roughly $1,600 to the costs of laparoscopy. When the overall cost of the robot itself was included, the added costs climbed to $3,200.

The Five Pillars and Robotic Surgery

While robotic surgery has demonstrated clear advantages, from significantly reduced patient recovery times to improved ergonomics for surgeons, the journey toward widespread adoption has been challenging, with significant barriers aligned closely with the Five Pillars of Healthcare Access.[15] Each of these pillars has historically impacted how robotic technologies were integrated into healthcare systems, presenting unique hurdles that had to be overcome.

Initially, **Physical Barriers** of robotic systems posed considerable challenges. The earliest robotic surgery devices were cumbersome and bulky, making them difficult to accommodate within smaller or remote healthcare facilities. Their size and weight restricted deployments primarily to large, urban medical centers that had the space and infrastructure necessary to house such sophisticated technology. For those who haven't had the experience of really seeing a surgical robot in action, it can be a bit intimidating. Imagine a Kraken hell bent on fixing you!

Financial Barriers compounded these challenges. Early robotic systems were extraordinarily costly, often running into millions of dollars of capital expenditure. Such prohibitive expenses limited their adoption to hospitals and healthcare networks in major urban areas with substantial funding and resources. This financial hurdle inherently excluded many rural hospitals and smaller clinics from embracing this innovative technology. Even though smaller, more cost-conscious robots are entering the market and large robotics companies now provide subscription-based pricing, the **Financial Barriers** remain.

Resistance also presented a notable barrier **(Cultural Barrier)**. Clinicians and surgeons initially viewed robotic surgery with

---

[15] Rehman S, Lopez PP, Donthi DN, et al. Ergonomic challenges of robotic surgery: a review. *Cureus*. 2023;15(10):e50415. Accessed March 24, 2025. https://pmc.ncbi.nlm.nih.gov/articles/PMC10784205/pdf/cureus-0015-000 00050415.pdf.

considerable skepticism. Many feared the technology would diminish the critical role of the human surgeon or even threaten their professional relevance altogether. This apprehension was widespread, and it took significant time and effort, along with successful clinical demonstrations, to begin shifting these perceptions. This is much like the fears many have today around AI taking over jobs, roles, and livelihoods.

Lastly, **Trust/Knowledge Barriers** played a crucial role. Surgeons often expressed distrust toward the precision, reliability, and safety of robotic systems, primarily due to limited formalized training programs and opportunities to become proficient with these new technologies. Without structured education and hands-on experience, surgeons were understandably cautious, slowing adoption rates as skepticism and resistance lingered.

Stories from the Field: Digital Barrier Spotlight[16]

Even when the robots are available and the surgeon is willing, time itself becomes a barrier. In a recent conversation with a surgeon working inside one of the nation's best known integrated health systems, I heard a surprising reflection. The surgeon, a specialist, did not dwell on the cost of robotics or the complexity of cases; instead, he highlighted the invisible time constraints that prevent specialists from meaningfully participating in robotic innovation.

"I only get a fixed amount of paid time off," he explained. "And unlike some of my general surgery colleagues, I don't do inpatient rounding or take daily call. That means I accrue less. So, when conferences or advanced training opportunities come up, I have to use up my vacation."

In other words, even among high-performing, full-time surgeons, access to professional development is unevenly distributed; based not on clinical need or surgical potential, but on how the job is structured. In a system designed for fairness, the unintended consequence is rigidity. While his organization offers support for innovation in theory, the realities of his clinical workload mean that ongoing robotic training, especially in emerging platforms, is something he must personally negotiate against his family time, personal recovery, or educational leave. That's not sustainable, nor scalable.

"I went into this role thinking I'd have more room to grow," he told me. "But the structure doesn't reward that. I feel like I have to justify every hour I want to spend learning something new."

---

[16] Anonymous surgeon; interview with Sarah Matt. April 15, 2025; personal communication.

This narrative reflects a deeper truth about the **Trust/Knowledge Barrier:** Adoption is not only about whether someone *wants* to learn, but whether the system creates space and incentives for that learning to happen. Time is a resource like any other, just as critical as hardware or funding. And when it's limited, innovation stalls.

Infrastructure: The Backbone of Robotic Surgery

Physical and **Digital Pillars:** Robotic surgery is transforming healthcare by overcoming traditional barriers to specialist care. However, this transformation is fundamentally dependent on robust physical and sophisticated digital infrastructure. Stable broadband connectivity, reliable electricity, interoperable digital systems, and well-maintained physical facilities are essential elements without which robotic surgery cannot reliably function. The reliance on these infrastructure pillars makes robotic surgery both powerful and vulnerable.

Even when the surgeon sits in the same operating room, every robotic platform behaves like a high-performance computer vision system. Multiple 3D high-definition video streams, fluorescence channels, and instrument telemetry must move instantly from the patient cart to the surgeon console. Simultaneously, the robot may retrieve preoperative images from the imaging system and write live data back to the EHR while anesthesia monitors and ancillary devices exchange commands in real time. These data flows demand stable bandwidth with low latency. If throughput drops, image quality deteriorates, instrument response lags, and the system can lose synchronization with vital imaging and monitoring feeds, jeopardizing precision and safety. Consequently, robust connectivity inside (not just outside) the hospital underpins every robotic case, whether the surgeon is across the room or across the ocean.[17]

Imagine the scenario of a surgical team mid-operation; reliable electricity is crucial because even a brief power outage could lead to equipment failure or procedure interruption, compromising patient safety. I remember operating during an earthquake back in the early 2000s, and while the electricity stayed on, I have never stitched someone up faster than I did that day!

---

[17] Schorp V, Giraud F, Pargätzi G, et al. A modular edge device network for surgery digitalization. *arXiv*. Published March 18, 2025. Accessed June 22, 2025. https://arxiv.org/abs/2503.14049.

Specialized software platforms serve as the invisible backbone of robotic surgery, coordinating complex interactions between robotic equipment and the rest of the surgery suite. Without such software, the intricate choreography required during robotic procedures would quickly become impossible to manage. Finally, the importance of physical facilities cannot be overlooked. Robotic surgical suites must be specially designed with adequate space to accommodate advanced robotic systems, proper ventilation to maintain air quality, and precise environmental controls to ensure that the equipment operates optimally, thus safeguarding patient health throughout the surgery.

To fully appreciate how infrastructure shapes robotic surgery's evolution, consider these significant historical milestones:

- **Early 2000s:** Introduction of robotic systems like da Vinci was limited primarily to resource-rich institutions due to high infrastructure requirements.

- **2010s:** Improvements in broadband and EHR systems facilitated broader integration of robotic surgery, especially in urban areas.

- **2020s:** The COVID-19 pandemic dramatically accelerated the adoption of telehealth and robotic surgery, simultaneously revealing critical infrastructure deficits in rural and underserved regions.

- **Future Outlook:** Infrastructure enhancements in underserved areas are recognized as essential for achieving patient-centered healthcare access and maximizing robotic surgery's potential.

This timeline highlights the essential alignment of healthcare innovation and infrastructural development, crucial for sustainable and balanced healthcare access in the United States. While we still struggle around the world to ensure adequate infrastructure for the latest surgical technologies, obtaining this foundation can mean getting more access to life-saving surgical procedures.

Broadband and Digital Connectivity: The Missing Link

Broadband connectivity is crucial not only for robotic surgery but also for telehealth, AI diagnostics, and remote patient monitoring. Households in non-metropolitan areas are significantly more likely to have limitations in device and high-speed Internet access compared to households in metropolitan areas. A 2023 study showed that

48.01 percent of rural households and 30.64 percent of urban household in the United States still lacked access to broadband Internet.[18] Similarly, just 65 percent of Native American households on tribal lands had broadband Internet compared to 99 percent of housing in U.S. urban areas, severely restricting their telehealth options.[19] Even in urban areas, low-income communities still face barriers due to broadband costs and digital literacy gaps.

A Tale of Two Countries: Rwanda and Japan[20]

Establishing robust infrastructure is critical when preparing a hospital system for advanced technologies such as robotic surgery, and this is paramount when we consider the complexities emerging for remote robotic procedures. Rwanda and Japan offer valuable contrasting examples of how infrastructure readiness influences digital healthcare transformation, providing insights relevant to robotic surgery adoption.

Rwanda's proactive infrastructure investment has significantly advanced healthcare delivery. The country prioritized expanding reliable electricity and widespread Internet access, creating a foundation supporting telemedicine and robotic surgery. Rwanda has aggressively implemented EHRs, with notable success in district hospitals. A recent study evaluating 257 health facilities across Rwanda reported that all 42 district hospitals had adopted EHRs, 83.3 percent specifically employing the OpenMRS platform. However, the adoption rate at smaller health centers remained modest at 33 percent, primarily due to challenges such as system maintenance issues and inconsistent Internet connectivity. These findings underscore Rwanda's commitment to digital transformation and highlight ongoing infrastructure development needs critical for future expansions of other advanced medical technologies.[21]

---

[18] Whitacre BE, Gallardo R, Grant A. Rural broadband availability and adoption: evidence, policy challenges, and options. *Health Aff (Millwood)*. 2023;42(1):123–131. Accessed March 24, 2025. https://pmc.ncbi.nlm.nih.gov/articles/PMC9827725/pdf/nihms-1851104.pdf.

[19] U.S. Department of the Interior, Bureau of Indian Affairs. *Expanding Broadband Access*. Published 2023. Accessed March 24, 2025. https://www.bia.gov/service/infrastructure/expanding-broadband-access.

[20] Nawrat A. Setting an example: Rwanda as a digital health success story. *Medical Device Network*. Published June 25, 2020.

[21] Muhire ML, Dushimimana E, Habineza H, et al. Assessment of the use of electronic medical records system and barriers in Rwanda. *Research Square*. Preprint posted online March 5, 2024. Accessed March 29, 2025. https://assets-eu.researchsquare.com/files/rs-4763866/v1/306eb0cc-5498-4756-bcc9-1ea5396fec6e.pdf?c=1724651484.

Conversely, Japan, despite its technological sophistication, has encountered significant challenges in achieving comprehensive EHR adoption across healthcare facilities. As of 2018, only 38.3 percent of Japanese hospitals had adopted EHRs, with adoption rates varying markedly by hospital size (62.5% in major hospitals, 21.7% in medium-sized hospitals, 9.1% in small hospitals, and 16.5% in clinics). Factors such as high implementation costs, fragmented hospital systems, and low digital literacy among older physicians contributed to this slower adoption rate. This fragmentation notably impeded interoperability, thereby slowing comprehensive digital healthcare integration, which could pose similar challenges to the adoption and effective implementation of robotic surgery programs.[22]

These contrasting experiences from Rwanda and Japan illustrate the diverse strategies and challenges associated with adopting complex digital healthcare technologies, such as robotic surgery.

Stories from the Field: Financial Pillar Spotlight in Ghana

Not every telemedicine initiative achieves lasting success, surgical or otherwise. Ghana's national telemedicine program, initially supported by the Novartis Foundation, illustrates the complexities involved in sustaining digital health innovation. The program significantly improved healthcare access, as nurses and doctors in centralized call centers effectively triaged rural patients. This system dramatically reduced unnecessary patient travel, enhancing emergency response times in remote regions.[23]

However, when donor funding was withdrawn, critical vulnerabilities surfaced. Ghana faced severe staffing shortages, which hampered continuous operations. Rural clinics struggled with unreliable electricity and Internet connectivity, fundamentally undermining the service's effectiveness. Additionally, the telemedicine initiative was inadequately integrated into Ghana's national healthcare infrastructure, ultimately contributing to its instability.

This Ghanaian experience underscores an essential lesson applicable to robotic surgery and advanced surgical technologies—without

---

[22] Kaneko M, Matsumoto M, Yano E. Factors associated with electronic health record use in primary care practices in Japan: a nationwide survey. *Interact J Med Res*. 2018;7(1):e11.

[23] Novartis Foundation. *Ghana Telemedicine*. Accessed March 30, 2025. https://www.novartisfoundation.org/past-programs/digital-health/ghana-telemedicine.

robust, stable infrastructure, sustainable funding, and governmental commitment, even the most promising digital health programs risk failure. Regions in rural India and across Sub-Saharan Africa face analogous hurdles, including unreliable electricity, limited broadband access, and insufficient healthcare facilities. These inadequacies reflect Ghana's experience, highlighting the critical need for global investment and strategic policy interventions to build the foundational infrastructure essential for the adoption of advanced surgical technologies and telemedicine services worldwide.[24,25]

Remote Robotic Surgery: Overcoming the Final Barrier

Remote robotic surgery, often called *telesurgery*, represents a radical evolution in healthcare delivery. By allowing surgical specialists to operate from distant locations, it removes the **Physical Barrier** entirely and democratizes access to complex procedures for patients wherever they live. Sustained adoption, however, hinges on resolving challenges across all Five Pillars of access, from reliable broadband and capital cost to licensure frameworks and patient confidence.

At the same time, telesurgery offers a 21st century answer to chronic surgical workforce shortages that have long plagued rural and underserved communities (see Diagram 5.4). Policy makers have tried to close these gaps for more than a century. Early legislation such as the Sheppard Towner Act of 1921 funded maternal and infant clinics, and the modern National Health Service Corps continues to use scholarships and loan repayment to steer clinicians toward shortage areas. Even so, many counties still lack surgeons in the United States. Telesurgery extends these historic efforts by delivering expert hands without requiring clinicians to relocate.

Currently, more than 77 million Americans live in federally designated healthcare professional shortage areas (HPSAs), according to the Health Resources and Services Administration.[26] Of the more

---

[24] Bervell B, Al-Samarraie H. A comparative review of mobile health and electronic health utilization in sub-Saharan African countries. *Health Policy Technol.* 2022;11(4):100792. Accessed March 30, 2025. https://pmc.ncbi.nlm.nih.gov/articles/PMC9761083/pdf/main.pdf.

[25] Bervell B, Al-Samarraie H. Barriers to the use of mobile health in improving healthcare services in Ghana: an analysis of existing literature. *Health Behav Policy Rev.* 2021;8(6):843–858. Accessed March 30, 2025. https://pmc.ncbi.nlm.nih.gov/articles/PMC8653215/pdf/HBE2-3-843.pdf.

[26] Health Resources and Services Administration. *Shortage Areas.* Health Workforce Data. Accessed March 29, 2025. https://data.hrsa.gov/topics/health-workforce/shortage-areas.

PILLAR OF ACCESS	IMPACT OF REMOTE ROBOTIC SURGERY
Physical Pillar 	Allows expert surgeons to conduct surgeries remotely, eliminating the need for patient travel and significantly improving access in rural or isolated regions.
Financial Pillar 	Reduces travel, lodging, and indirect patient expenses associated with obtaining specialized care, improving affordability for both patients and healthcare systems.
Cultural Pillar 	Ensures patients have access to culturally appropriate specialist care, regardless of geographic location, supporting trust-building and patient engagement.
Digital Pillar 	Encourages infrastructure investment in reliable, interoperable digital platforms and high-speed Internet, critical for real-time surgical procedures.
Trust/Knowledge Pillar 	Enhances patient and clinician confidence through demonstrated success in remote interactions, transparent communication, and high-quality clinical outcomes.

Diagram 5-4: Impact of Remote Robotic Surgery on Healthcare Access.

than 7,200 federally designated health professional shortage areas, three out of five are in rural regions. And while 20 percent of the U.S. population lives in rural communities, only 11 percent of physicians practice in such areas.[27] While technological innovations like AI and

---

telemedicine can support healthcare professionals, they cannot entirely replace human providers. Without addressing underlying workforce shortages, these advanced technologies risk becoming underutilized.

By confronting and overcoming historical challenges aligned with the five pillars, surgery continues to redefine healthcare delivery. Continued advancements in technology, targeted investments, improved training protocols, robust digital infrastructure, and growing cultural acceptance are collectively paving the way toward accessible healthcare for all.

Financial Barrier: Remote robotic surgery requires significant investment in technology, infrastructure, and ongoing operational costs. While the potential savings from improved patient outcomes and reduced hospital stays are considerable, the initial and ongoing expenses can be substantial. Traditional reimbursement models often lag technological innovation, making it challenging for healthcare providers to adopt remote robotic surgery without clear financial incentives or support from insurance companies and governments. Sustainable funding models, such as bundled payments or value-based reimbursement, may help overcome these economic hurdles by incentivizing efficient, high-quality remote surgical care. With the potential to have two different locations and perhaps two surgeons (one on each side) as this modality matures, reimbursing all these personnel for their activities may be a challenge. Additionally, every country or state may have different views on how reimbursement is implemented, approved, or paid out, which may stall adoption based on region.

Physical Barrier: Unlike traditional robotic surgery confined to a single operating room, remote robotic surgery requires a reliable physical and technological infrastructure to support real-time communication and precision movements across distances. High-speed Internet, low-latency connections, and dedicated bandwidth are crucial for ensuring uninterrupted, secure, and safe surgical procedures. Unfortunately, many rural or underserved regions lack sufficient broadband or telecommunications infrastructure, severely limiting the deployment of remote robotic surgery in those areas. Investments in infrastructure, such as broadband expansion, are essential for widespread adoption.

Cultural Barrier: Cultural attitudes toward technology, particularly the idea of surgery performed remotely, present significant hurdles. Patients may be hesitant to trust surgical care delivered from afar, perceiving remote procedures as impersonal or inferior to traditional

March 29, 2025. https://www.aamc.org/news/attracting-next-generation-physicians-rural-medicine.

face-to-face surgery. This was a barrier when telemedicine for non-surgical visits first emerged. Surgeons, too, may initially resist remote robotic surgery due to concerns about professional autonomy or discomfort with new practices. This also plays into liability, and who is "responsible" for the patient, requiring clinical guidelines that clearly articulate the role of members of the care team regardless of location. Addressing these **Cultural Barriers** requires active education and demonstration of remote robotic surgery's benefits, safety, and effectiveness, alongside the involvement of trusted local healthcare providers who champion and advocate for the technology.

Digital Barrier: The digital demands of remote robotic surgery extend beyond basic connectivity. The integration of remote robotic platforms into existing hospital information systems, EHRs, and telehealth infrastructures demands sophisticated interoperability and seamless data exchange. Without standardized protocols and robust digital infrastructure, remote surgery cannot achieve its potential. Hospitals and remote clinics must prioritize interoperability standards such as FHIR (Fast Healthcare Interoperability Resources), DICOM (Digital Imaging and Communications in Medicine), and HL7 (Health Level Seven International). Equally essential is rigorous training and support for clinical teams to ensure these systems enhance rather than disrupt care workflows.

Trust/Knowledge Barrier: Perhaps the most sensitive and complex barrier to overcome is establishing trust and confidence in remote robotic surgical platforms. Surgeons must feel fully confident in the precision, reliability, and responsiveness of remote technologies before incorporating them into clinical practice. Similarly, patients must trust the technology and the surgeons performing their procedures remotely. Building this trust requires comprehensive training programs, clear demonstrations of safety and efficacy, and consistent, transparent communication with patients and healthcare providers about risks, benefits, and outcomes. Ongoing education and clinical mentoring—especially leveraging tele-mentoring capabilities inherent in robotic systems—can help bridge knowledge gaps, alleviate anxiety, and foster greater acceptance and clinical confidence.

By systematically addressing these interconnected barriers across the five pillars, remote robotic surgery holds the potential to radically transform healthcare delivery. The future of remote surgical care hinges not merely on technological advancements but also on thoughtfully designed healthcare systems that prioritize inclusivity, sustainability, and patient-centered care.

Champions and Training: Human Engines of Robotic Adoption

Local "champions" are the critical human element behind successful digital and robotic surgical rollouts. These individuals, often experienced surgeons or clinicians with credibility among peers, don't simply introduce new technology; they actively transform institutional cultures around healthcare innovation. Their advocacy is crucial in overcoming skepticism and resistance to change, particularly in settings unfamiliar or uncomfortable with technology-driven care.

In the realm of robotic and especially remote robotic surgery, champions play a pivotal role. Surgeon champions not only demonstrate the clinical benefits of robotic platforms, such as precision, minimally invasive techniques, and improved ergonomics, but also actively dispel myths and ease fears among colleagues and patients alike. Early adopters of robotic prostatectomy became advocates who reshaped urology. They made robotic-assisted, minimally invasive prostatectomy the new standard of care despite initial resistance and skepticism. These pioneers demonstrated improved patient outcomes and reduced recovery times. Ultimately, they influenced both professional opinion and patient demand in favor of robotic approaches. Most recently, Dr. Vipul Patel of Orlando's Advent Health carried that spirit of advocacy to a global stage when he completed the first, transcontinental remote robotic prostatectomy from Orlando to Angola. Reflecting on the milestone, he remarked, "It was a small step for a surgeon, but a huge leap for healthcare."[28]

The role of "champion" becomes even more crucial in remote robotic surgery, where trust in both the technology and the remote surgeon must overcome significant **Cultural Barriers**. Here, champions like Dr. Dennis Fowler, a surgical pioneer who spearheaded minimally invasive technique adoption at multiple institutions, have demonstrated the necessity of both cultural engagement and clear communication to foster acceptance. Fowler's approach involves systematically addressing colleagues' concerns, sharing success stories, and continuously reinforcing the real clinical value of robotic surgery. His experience illustrates that adoption accelerates when trusted clinical leaders become vocal supporters, embedding the technology into

---

[28] Benadjaoud Y, Serratos O, Mendelsohn M, Miller M. Exclusive look at groundbreaking remote robotic surgery: patient was in Africa; doctor was in Florida. *ABC News*. Published June 17, 2025. Accessed June 22, 2025. https://abcnews.go.com/Health/exclusive-groundbreaking-remote-robotic-surgery-patient-africa-doctor/story?id=122946197.

the professional culture of their institutions. In rural or underserved regions, where cultural resistance can be especially pronounced, a single trusted physician champion can be transformative, translating initial curiosity into sustained adoption.

Structured Training: Building Knowledge and Confidence

Robotic and remote robotic surgery cannot thrive on enthusiastic advocacy alone. Comprehensive, structured training programs are essential, systematically designed to build surgical skill, clinical knowledge, and trust in robotic platforms. One globally recognized model is the ORSI Academy in Belgium, a premier international training facility dedicated to robotic surgery. ORSI provides structured, iterative, hands-on learning experiences that cater to surgeons at various levels of proficiency. Similarly, IRCAD with its HQ in Stroudsburg, France conducts basic and advanced hands-on, as well as web-based, courses at their nine global locations. Trainees benefit from direct mentorship, virtual simulations, and collaborative workshops; all designed explicitly to enhance technical skills and build the surgeon's confidence in robotic technology. These proven training methodologies ensure that surgeons worldwide are uniformly prepared, resulting in safer and more effective patient outcomes and broader acceptance across diverse healthcare contexts.

The importance of structured training escalates in remote robotic surgery scenarios, where technical complexity and psychological factors are both heightened. Robust simulation training, tele-mentoring, and dual-console robotic systems further enhance knowledge transfer and skill development, potentially providing a safety net for surgeons learning complex procedures remotely. This training can reassure surgeons and patients alike, reducing anxieties about remotely conducted operations and building broader trust in these groundbreaking interventions.

 Next Shift Quick Wins

1. **Run a Five-Minute "Robot Readiness" Drill at First Case Huddle:** Confirm system self-test passed, bedside assist trained, rescue lap instruments opened.

2. **Publish Yesterday's Robotic Case Length and Conversion Rate on the OR Whiteboard:** Transparency drives peer-to-peer coaching.

3. **Add One Question to the Patient Discharge Call Script:** "Did the robot make you feel safer, the same, or less safe?" Feed results to Quality every Monday.

Stories from the Field: Financial Barrier Spotlight

In rural Alabama, the ongoing decline of local healthcare services has significantly impacted surgical access. **Financial Barriers** have caused the closure of healthcare facilities, leading to a scarcity of surgeons and medical professionals. Patients must now travel considerable distances for even routine surgeries, exacerbating health disparities and resulting in delayed interventions and poorer outcomes. The absence of robotic surgical technology in these areas further limits local capabilities, highlighting the urgent need for innovative funding solutions to sustain rural surgical services.[29]

Ensuring robust financial frameworks for robotic surgery is paramount. Early robotics faced considerable hurdles due to high acquisition and operational costs. Yet, as robotic surgery has demonstrated clear advantages—including reduced complications, shorter hospital stays, quicker patient recoveries, and fewer readmissions—health systems and policymakers have begun recognizing these long-term financial benefits.

Innovative funding strategies have become critical to expanding robotic access. Public-private partnerships (PPPs) have emerged globally as a valuable model. These partnerships enable hospitals, government entities, and private companies to share investment burdens and benefits, reducing initial financial barriers. Bulk purchasing agreements and shared service models further enable smaller or rural hospitals to adopt robotic platforms by collectively negotiating better pricing and distribution of maintenance costs, making advanced surgical technology affordable beyond traditional, wealthy metropolitan areas.

---

[29] Sheets C. *The Long Decline: Health Care Access Grows Difficult in Shrinking Rural Communities*. Alabama Reflector. Published January 18, 2024. Accessed March 29, 2025. https://alabamareflector.com/2024/01/18/the-long-decline-health-care-access-grows-difficult-in-shrinking-rural-communities/?utm_source=chatgpt.com.

Stories from the Field: Financial Pillar Spotlight in Japan[30]

Japan, recognized globally for its technological advancement and aging population, has rapidly adopted robotic surgery. By 2023, over 570 da Vinci surgical systems were operational nationwide. While this adoption has marked significant progress, it has simultaneously highlighted ongoing healthcare access gaps.

Physical Barrier: Robotic surgical systems have proliferated primarily in urban hospitals, which perform significantly more robot-assisted radical prostatectomy (RALP) procedures (937 procedures on average) compared to regional hospitals (195.5 procedures). This highlights a clear geographic divide.

Financial Barrier: This barrier has been partially addressed through regulatory alignment and national insurance coverage introduced in 2012. However, sustainable reimbursement structures favor urban centers due to higher patient volumes, leaving regional hospitals financially disadvantaged.

Cultural Barrier: This barrier remains a challenge, as the cultural trust in face-to-face interactions persists strongly among the older Japanese population. This portion of the population is the largest percentage, as Japan has a super-aging population. Potential remote surgical procedures, despite technological sophistication, require considerable effort to build patient trust and acceptance.

Digital Barrier: This barrier varies across regions. While the infrastructure in urban hospitals supports advanced robotics and telemedicine, rural facilities often struggle with infrastructure limitations, hindering their ability to utilize such technologies effectively.

Trust/Knowledge Barrier: This barrier is influenced significantly by the availability of trained specialists. Regional hospitals face challenges in providing sufficient procedural volume experience to local surgeons, impacting their ability to gain certification and patient trust. Remote support technology has begun bridging this gap by allowing regional surgeons real-time guidance from urban-based specialists.

---

[30] Hara K, Kanda M, Kuwabara H, Kobayashi Y, Inoue T. Current status analysis of the prevalence and regional disparities of robot-assisted laparoscopic prostatectomy in Japan using diagnosis procedure combination data. *Sci Rep.* 2024;14:24823. doi:10.1038/s41598-024-75837-9.

Japan's recent initiatives offer a compelling case study: national investment into robotic surgical programs exemplifies how proactive, centralized financial policies can dramatically increase access. Japan systematically supported hospital adoption of robotic technology through public funding and strategic partnerships, resulting in broader geographic distribution of surgical robots, improved patient outcomes, and a more cost-effective healthcare delivery system overall. Cost-effectiveness analyses conducted during these initiatives consistently highlighted reduced hospital stays and fewer surgical complications, reinforcing both clinical and economic rationales for ongoing investment in robotic platforms. Addressing these disparities will require targeted investment in training programs, balanced distribution of specialists, robust infrastructure improvements, and focused efforts to enhance digital literacy and cultural acceptance, ensuring a cohesive healthcare system that benefits all geographic regions.

Regulatory Framework: Navigating Trust, Knowledge, and Digital Infrastructure

A supportive regulatory environment is just as essential as financial investment. Regulatory frameworks directly influence the speed and extent of technological adoption by shaping the trust among patients and clinicians **(Trust/Knowledge Pillar)**, standardizing digital practices, and providing clear, consistent reimbursement guidelines. Regulatory clarity, particularly concerning FDA approvals in the United States, and insurance reimbursement are vital to the success and widespread adoption of robotic surgery. Surgeons and healthcare providers require certainty around which procedures, platforms, and telehealth components will receive coverage and approval, thereby minimizing financial risk and incentivizing adoption. Like any business, healthcare systems depend on revenue generation to pay for their employees (surgeons), facilities (hospitals) and materials (surgical instruments and medical supplies). If you can't plan for these elements, it's very difficult to include robotic surgery as a part of your business.

Robotic surgical systems, particularly those capable of remote operation, introduce unique regulatory complexities. These include compliance with rigorous medical device standards, cybersecurity safeguards, and robust patient privacy protections, alongside assurances of clinical safety and effectiveness. International regulatory alignment, like mutual recognition of device approvals or coordinated telehealth policies, can significantly accelerate the global adoption of robotic surgery, reducing barriers and encouraging broader implementation.

The COVID-19 pandemic provided an unprecedented case study in regulatory flexibility and responsiveness. Regulatory changes and temporary policy relaxations, like expanded telehealth reimbursements and streamlined device approvals, quickly boosted remote healthcare technology adoption. This period demonstrated the significant positive impact that regulatory agility could have on healthcare innovation adoption, especially when paired with clear, consistent reimbursement mechanisms.

Financial sustainability and regulatory frameworks are inherently interconnected. Successful adoption of robotic and remote robotic surgery depends heavily on clear, aligned policy guidance and economically sustainable frameworks. Financial incentives must be paired with supportive regulatory structures, such as insurance reimbursement policies for robotic procedures, funding support for infrastructure enhancements (notably broadband access critical for telesurgery), and regulatory clarity around device standards and approvals.

 How-To: Planning Your Remote-Ready Playbook

1. **Map the Current State:** List every robot-enabled procedure, average case volume, unplanned conversion percentage, and presence of specialty champions. Determine the goal of "remote" at your organization and what use cases you will address.

2. **Gap-Assess Against Remote-Surgery Requirements:** Identify where technical gaps in fiber coverage, dual consoles, or other hardware exist.

3. **Secure Early Partner Commitments:** Line up a fiber vendor, a malpractice carrier willing to cover tele-presence, and a surgical device firm ready for an FDA investigational device exemption.

4. **Prototype a "Dual Console Shadow" Workflow:** Run three live cases with the mentor at a second console in the same room to harden escalation and comms protocols.

5. **Draft an IRB Concept Sheet:** Include data capture on network performance, near-misses, and patient-reported trust to accelerate future submissions.

The Human Touch: Training

Dr. Fowler emphasizes that robotic surgery is not an individual endeavor but a highly orchestrated team sport. His insights, drawn from years overseeing perioperative services, show how essential specialized training and consistent teamwork are for successful surgical programs. Fowler observes that the most effective surgery teams specialized in specific disciplines, underscoring that "the best care comes when the same team does the same operation hundreds or thousands of times."[31] However, hospitals that insist on generalist teams struggle, experiencing higher stress levels and increased error rates. Fowler's implementation of structured team training at several institutions—focusing on closed-loop communication, situational awareness, and clear escalation protocols—marks a turning point. It demonstrates that adopting any surgical innovation isn't just technological; it's fundamentally about transforming hospital culture and communication to support new tools safely and effectively.

> The best care comes when the same team does the same operation hundreds or thousands of times.

Fact Check

- Robotic-assisted surgery reduces postoperative complications by approximately 30 to 50 percent compared to traditional surgical methods, especially in urologic and gynecologic procedures.[32]

- Patients undergoing robotic procedures typically experience 20 to 30 percent faster recovery and shorter hospital stays relative to those treated with conventional surgery.[33,34]

---

[31] Fowler D. Interview with Sarah Matt, MD. March 21, 2025.
[32] Zaninotto P, Meghea CI, Abazari S, et al. Advancements in robotic surgery: a comprehensive overview of current utilizations and upcoming frontiers. *Cureus.* 2023;15(8):e43621. Accessed March 26, 2025. https://www.ncbi.nlm.nih.gov/pmc/articles/PMC10471642.
[33] Yang GZ, Cambias J, Cleary K, et al. Medical robotics—regulatory, ethical, and clinical challenges. *Sci Robot.* 2022;7(66):eabo7118. Accessed March 26, 2025. https://www.science.org/doi/10.1126/scirobotics.abo7118.
[34] EIT Health. *Robots in Medicine: The Robotics Start-ups Making a Difference.* European Institute of Innovation and Technology Health. Published 2020. Accessed March 26, 2025. https://eithealth.eu/news-article/robots-in-medicine-the-robotics-start-ups-making-a-difference.

Global Collaboration: Expanding Robotic Surgery's Impact

Robotic and remote robotic surgery represent transformative innovations, offering remarkable opportunities to address global surgical disparities. Fully realizing this potential demands comprehensive, multinational collaboration. Successful international initiatives highlight how coordinated efforts can bridge all the access-related barriers, ultimately scaling robotic surgery's impact globally. The global healthcare landscape is diverse; each region brings distinct challenges across the Five Pillars. Multinational collaborations effectively respond to these unique barriers by sharing resources, knowledge, and expertise.

Physical Pillar: Geographical isolation remains a considerable barrier worldwide, particularly in low-resource and rural settings. Countries such as Singapore and Japan exemplify how strategic global partnerships can mitigate these barriers. In a groundbreaking telesurgery trial, clinician-scientists from Singapore's National University Hospital and Japan's Fujita Health University demonstrated the feasibility of long-distance robotic surgery across more than 5,000 kilometers, offering a model for extending care without relocating patients.[35]

Cultural Pillar: Cultural alignment plays a critical role in the successful adoption of surgical innovation. In global robotic surgery efforts, outcomes improve when programs are developed in partnership with local healthcare providers who reflect and respect the cultural context of their communities. Rather than imposing foreign systems, these initiatives embed robotic surgery within existing care models, allowing trusted local surgeons to lead care delivery and bridge the gap between advanced technology and patient expectations. This approach, rooted in cultural humility and community engagement, has proven essential in reducing resistance and fostering trust in robotic surgical platforms.[36]

---

[35] National University Health System. *Robotic Telesurgery Trial Offers Glimpse into the Future of Healthcare.* Published 2024. Accessed March 30, 2025. https://nuhsplus.edu.sg/article/robotic-telesurgery-trial-offers-glimpse-into-the-future-of-healthcare.

[36] Meara JG, Leather AJM, Hagander L, et al. Global Surgery 2030: evidence and solutions for achieving health, welfare, and economic development. *Lancet.* 2015;386(9993):569–624. doi:10.1016/S0140-6736(15)60160-X.

Digital Barrier: Limited broadband infrastructure, cybersecurity concerns, and interoperability challenges significantly hinder the scalability of remote surgical interventions in regions like Sub-Saharan Africa and Southeast Asia. In Sub-Saharan Africa, the limited availability of low-cost backbone network capacity constrains broadband connectivity development, impacting the feasibility of telemedicine services.[37] Similarly, in Southeast Asia, efforts to extend broadband connectivity are ongoing, with tailored recommendations being developed to improve access and support digital health initiatives.[38] Addressing these infrastructural deficits through targeted investments and policy reforms is crucial to enabling remote surgical procedures at scale.

Trust/Knowledge Pillar: Surgeons and patients alike require familiarity and trust in robotic systems. Global centers like IRCAD (France) and ORSI Academy (Belgium) offer structured training in robotic techniques. Their models combining in-person simulation, virtual mentoring, and stepwise skill-building can serve as blueprints for global upskilling.[39]

Financial Pillar: The high acquisition and maintenance costs of robotic surgical systems present significant financial barriers to widespread adoption. Capital expenses can reach at least $1.5 million, with recurring annual costs around $100,000.[40] In Japan, the government has initiated efforts to decrease regulations and ensure funding to develop new surgical robotic systems. The New Energy and Industrial Technology Development Organization (NEDO) allocated $42 million within a $138 million framework to support the development

---

[37] Williams MDJ. Advancing the development of backbone networks in Sub-Saharan Africa. In: *Information and Communications for Development 2009: Extending Reach and Increasing Impact*. World Bank; 2009:85–98. https://thedocs.world bank.org/en/doc/761731434649062983-0190022009/original/IC 4D2009Chapter4.pdf.

[38] Organisation for Economic Co-operation and Development (OECD). *Extending Broadband Connectivity in Southeast Asia*. OECD Publishing; 2023. Accessed March 30, 2025. https://www.oecd.org/publications/extending-broadband-connectivity-in-southeast-asia-b8920f6d-en.htm.

[39] IRCAD France, Orsi Academy Belgium. *Robotic Surgery Training Programs*. Accessed March 30, 2025. https://www.ircad.fr and https://www.orsi-online.com.

[40] Burke J, Gnanaraj J, Dhanda J, et al. Robotic surgery in low- and middle-income countries. *Bull R Coll Surg Engl*. 2024;106(3):54–56. Accessed March 30, 2025. https://publishing.rcseng.ac.uk/doi/10.1308/rcsbull.2024.54.

of surgical robots, including regulatory reforms to facilitate clinical trials.[41] Similarly, in Singapore, the healthcare system benefits from strong public-private collaborations, contributing to high efficiency and cost control.[42] These initiatives demonstrate how strategic investments and collaborations can mitigate financial barriers, making advanced surgical technologies more accessible and promoting their broader adoption.

Addressing Challenges in Low and Middle-Income Countries (LMICs)[43]

Despite successful collaborations, LMICs still face unique, persistent barriers. Centralization strategies that work in high-income regions can inadvertently exacerbate imbalance in LMICs due to inadequate transportation infrastructure. Furthermore, the initial robotic system costs averaging 1.3 million USD per unit, plus ongoing maintenance, remains daunting **(Financial Barrier)**. To overcome these challenges, LMICs require structured international partnerships focused on sustainable infrastructure investments, community-oriented training, and culturally sensitive approaches that promote local ownership and clinical autonomy.

Furthermore, ethical considerations surrounding "medical imperialism" and reliance on "black box" AI algorithms necessitate careful navigation. International initiatives must emphasize local capacity-building, transparency, and cultural humility to sustainably integrate robotic surgery into LMIC healthcare systems. Proposals such as establishing an "international robotic surgical corps" can facilitate long-term, sustainable impact by fostering even-handed

---

[41] Japan's new robotics push: funding and deregulation. *The Robot Report*. Published September 2014. Accessed March 30, 2025. https://www.therobotreport. com/japans-new-robotics-push-funding-and-deregulation.

[42] *Singapore's Healthcare Blueprint: Efficiency, Innovation, and the Future of Medical Excellence*. MedTech Spectrum. Published March 2025. Accessed March 30, 2025. https://medtechspectrum.com/analysis/24/23979/singapores-healthcare-blueprint-efficiency-innovation-and-the-future-of-medical-excellence.html.

[43] Mehta A, Cheng Ng J, Andrew Awuah W, et al. Embracing robotic surgery in low- and middle-income countries: potential benefits, challenges, and scope in the future. *Ann Med Surg (Lond)*. 2022;84:104803. doi:10.1016/j.amsu.2022.104803. PMID: 36582867; PMCID: PMC9793116.

partnerships and comprehensive skill transfers, thereby directly empowering local healthcare providers.

Next Shift Quick Wins: Two-Week Tune-Ups for Safer Surgery Today*

Win	Time	Owner
Conduct a 15-minute "network sanity" drill; ping test and document latency before the first robotic case.	Day 1	Biomed Tech
Load a "what-if" power-loss scenario into the simulation cart and rehearse bedside takeover with the whole team.	Week 1	Charge Nurse
Add a second overhead camera and stream it to a secure team tablet so the anesthesia lead can see the console view in real time.	Week 2	OR Manager

Why now? Remote telesurgery is still limited to early clinical trials in the United States, and special cases around the world. These wins tighten onsite safety today while building habits that will matter when distance disappears.

Looking forward, sustainable scaling of robotic surgery requires a strategic balance across all five pillars. It necessitates ongoing investment in access through remote capabilities **(Physical Pillar)**, robust planning **(Financial Pillar)**, active engagement **(Cultural Pillar)**, continuous infrastructure enhancement **(Digital Pillar)**, and unwavering dedication to trust-building through education and transparency **(Trust/Knowledge Pillar)**.

Imagine a future where healthcare outcomes are entirely independent of geographic constraints. A future where every patient, regardless of location, receives timely, high-quality surgical care. Robotic surgery integrated seamlessly with telehealth can realize this vision. A surgeon in New York can perform life-saving operations on patients in rural Africa, Appalachia, the north country of New York state, or the remote Pacific islands, erasing historical gaps in surgical access. This is not merely aspirational; it is achievable. We stand at the threshold of this new reality, driven by lessons learned, strategic integration across pillars, and a commitment to making healthcare universally accessible. Geography will no longer dictate healthcare outcomes; robotic surgery is making that future our reality today.

Startup Builder's Box: Tools to Map Your Remote-Surgery White Space

DISCOVERY STEP	KEY QUESTIONS TO ANSWER	TOOL OR METRIC TO USE	EARLY PROOF YOU ARE ON THE RIGHT TRACK
1. Map the pain	Where do surgeons, nurses, and patients lose time; money; or confidence during today's robot cases?	One-day shadow map; 20 rapid-fire interviews; cost of delay worksheet	A single slide quantifying the top friction in dollars and minutes.
2. Scan the guardrails	What will FDA reviewers require for a novel network or workflow add-on?	Q-submission outline and feedback log	Written FDA notes that confirm your add-on can ride the existing robot clearance pathway.
3. Stress-test the tech	Does latency, jitter, or cyber-risk break your concept?	Bench rig with latency emulator; threat list; daily scorecard	Demonstration video that keeps task error below 2% at variable millisecond delay. (Because nobody wants their surgeon lagging like bad hotel Wi-Fi.)
4. Validate desirability	Will a real hospital pay for this fix before the big vendors move?	Lean canvas pricing game; letter-of-intent template	One signed intent letter or pilot agreement that defines success metrics and payment trigger.
5. Design the trust story	How will patients know a remote expert is as safe as a local one?	30-second animated consent; five-item teach-back quiz	90% correct teach-back rate in a hallway test with 20 patients.

Leaders' End-of-Chapter Action Checklist: Chapter 5: "The Robotics Revolution"

LEADER	HIGH-IMPACT ACTION TO STRENGTHEN SURGICAL ACCESS AND READINESS
❑ Board Director	Approve a five-year capital roadmap that replaces aging robotic platforms on a fixed schedule and prepares for needed infrastructure for remote capabilities
❑ Chief Executive Officer	Establish a Robotic Steering Council within 30 days; set a target of 85% on-time starts for the first case of robotic surgery each day and review results monthly
❑ Chief Information Officer	Install a redundant fiber pathway to every robotic operating room; determine organizational KPIs to be reported on Day 1 of remote surgery
❑ Chief Medical Officer	Add a robotic proficiency module to ongoing professional practice evaluation; mandate peer review of the first 20 cases for every newly credentialed surgeon with robotics as part of their practice
❑ VP Clinical Engineering & Biomed	Launch a predictive maintenance dashboard that pulls robot telemetry every night; achieve 99.8% equipment availability each quarter
❑ VP Supply Chain	Secure a consignment program for single use robotic instruments; cut on-hand inventory 25% and redeploy the freed capital to fund after-hours open simulation lab access
❑ OR Nurse Manager	Run a 10-minute emergency undocking drill every Wednesday before first incision; log team completion and debrief any errors
❑ Simulation Program Director	Provide 24/7 access to a portable robotic simulator cart; email a monthly leaderboard that shows practice hours for each resident and attending
❑ Patient Safety Officer	Embed a two-step Dock and Verify checklist in the EHR; analyze near-miss reports quarterly and publish lessons learned
❑ Director of Snacks & Morale	Celebrate every one hundredth robotic case with robot arm shaped cookies and decaf in the lounge; happy teams spot hazards sooner and keep patients safer

Intelligence Without Borders

The Intelligence Gap We Can Close

Artificial intelligence (AI) is not a miracle cure. It will not fix healthcare with a snap of its fingers. But it *can* expand access, if we build it right. The global AI-in-healthcare market reached US $19.27 billion in 2023 and is expected to grow at a CAGR of 38.5 percent from 2024 to 2030 (see Diagram 6.1).[1] Much of the public conversation about AI in healthcare leans heavily to either utopian or dystopian. One side promises that algorithms will diagnose rare cancers faster than any doctor can. The other warns that biased models will quietly entrench existing disparities while hiding behind glossy dashboards. Both contain truth. But neither tells the full story. Underneath the buzzwords, something quieter is happening. In pockets of real-world practice—from rural Nigeria to home health visits in the UK to diabetic retinopathy screening in underserved U.S. clinics—AI is already improving care. These aren't moonshot tech demos. They're tools, embedded in workflows, shaped by clinical need, and driven by local trust (**Trust/Knowledge Pillar**).

---

[1] Grand View Research, Inc. *Artificial Intelligence (AI) in Healthcare Market Size, Share & Growth Report, 2030*. Grand View Research, Inc.; 2024 Apr. Accessed April 12, 2025. https://www.grandviewresearch.com/industry-analysis/artificial-intelligence-ai-healthcare-market.

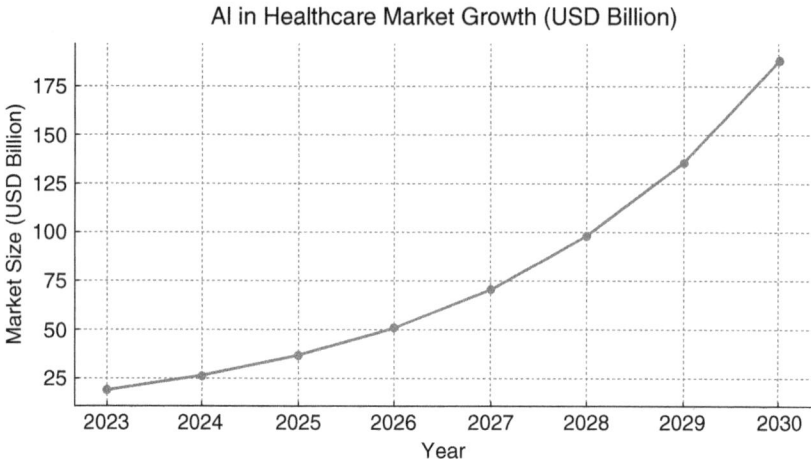

Diagram 6-1: Global AI Healthcare Market Growth, 2023–2030 (Projected).[2]

This chapter is not about hype. It's about possibility. Specifically, the possibility that we can use AI to *close* healthcare access gaps, not widen them. Not by replacing humans, but by augmenting them. Not by scaling broken systems, but by redesigning them around what patients and providers actually need. This chapter looks at how AI interacts with the Five Pillars of Access and where it succeeds or fails in practice (see Diagram 6.2). It explores where the tools are helping real patients right now, where they're hurting, and what it will take to build intelligent systems that serve *everyone*. The technology is already here. The intelligence gap, the real one, is how we choose to use it.

The Five Pillars Meet Their Match

Artificial intelligence is often touted as healthcare's next great leap forward—an all-seeing, always-learning engine poised to detect disease earlier, diagnose it more accurately, and deliver care more efficiently. But AI does not thrive in a vacuum. Like every other health-care innovation, its impact is defined not just by what it *can* do, but by where and how it is used.

To understand that, let's return to a foundational lens: the Five Pillars of Access. These are not just the barriers that AI must overcome; they

---

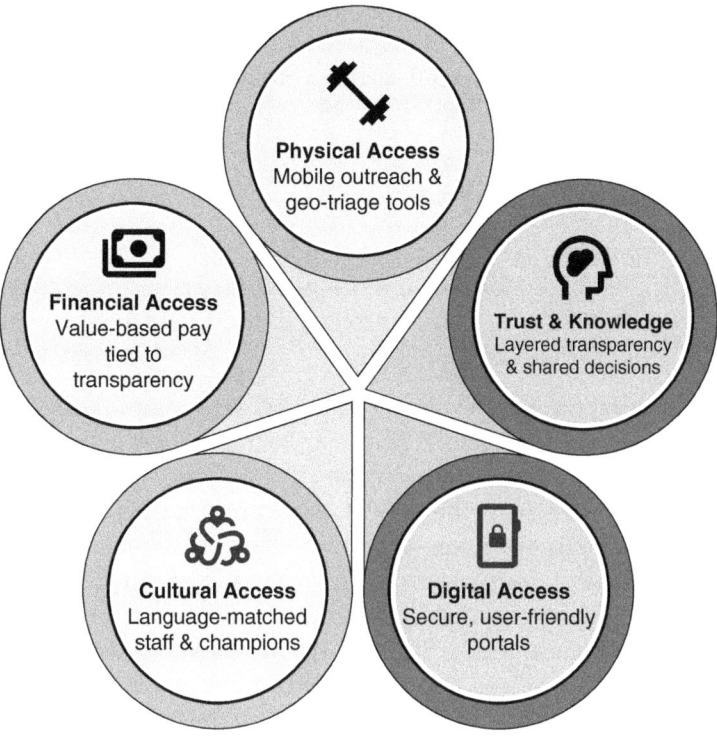

Diagram 6-2: Pillar Spotlight.

are the criteria any real-world healthcare tool must meet if it hopes to expand access universally.

- **Physical Pillar:** Can the solution reach people where they are?
- **Financial Pillar:** Is it affordable to those who need it most?
- **Cultural Pillar:** Does it align with the beliefs, language, and social norms of its users?
- **Digital Pillar:** Can it function within existing infrastructure? Does it require connectivity that isn't there?
- **Trust/Knowledge Pillar:** Do patients and clinicians trust it enough to use it? Do they understand how?

In the context of AI, these pillars translate into tangible design and implementation requirements. That includes regulatory clarity (for liability and approval), sustainable business models (for long-term updates and retraining), interoperable systems (to make use of fragmented data), cultural alignment (in both interface and output), and intentional governance structures that foster trust across diverse populations. When we look at the field today, it's clear: AI is not yet a system-level equalizer, but it *can* be. In fact, we already see moments where AI meets these five pillars head on. Not by sidestepping complexity, but by designing for it.

Stumbles and Successes

For every stumble on the path toward adopting AI, there is an equally interesting success story. At a more granular level, there are great situations that show how to overcome each barrier to care or inadvertently build these barriers higher.

Physical Pillar: In Nigeria, the startup Ubenwa exemplifies how AI is being harnessed to overcome **Physical Barriers** and diagnostic limitations in low-resource environments. By applying signal processing and machine learning to the cries of newborns, Ubenwa enables early detection of birth asphyxia, a leading cause of neonatal mortality, without requiring invasive procedures or high-cost clinical infrastructure.[3] This approach transforms simple audio data into life-saving diagnostic signals, offering frontline health workers a scalable tool in regions where advanced neonatal care is often unavailable.

Meanwhile, another AI-driven breakthrough is reshaping diabetic retinopathy (DR) screening in developing nations, where rural populations face a chronic shortage of ophthalmologists and diagnostic equipment. AI-based screening tools have demonstrated high sensitivity and specificity for DR detection. This has enabled swift, accurate evaluations, even in facilities lacking onsite specialists.[4,5] This innovation relieves pressure on overburdened referral centers and mitigates the diagnostic delays that so often accompany geographic isolation.

Together, these examples illustrate how AI, when embedded in tools designed for physical and systemic constraints, can function as an equalizer in healthcare delivery, not just by extending care into hard-to-reach places, but also by reconfiguring what *counts* as infrastructure in the first place.

---

[3] Nair M, Ameyaw M, Abu EK, et al. Exploring the impact of artificial intelligence on global health and enhancing healthcare in developing nations. *Ann Med Surg (Lond)*. 2024;85:1221–1226. doi:10.1016/j.amsu.2024.122126. https://www. ncbi.nlm.nih.gov/pmc/articles/PMC11010755/.
[4] Ibid.
[5] Dey AK, Walia P, Somvanshi G, et al. Artificial intelligence–driven diabetic retinopathy screening: multicentric validation of AIDRSS in India. *Preprint*. arXiv; published January 10, 2025. Accessed June 27, 2025. https://arxiv.org/ abs/2501.05826.

In stroke care, time is oxygen. The difference between disability and recovery often comes down to minutes and rural hospitals rarely have the staff or resources to handle high-stakes triage alone. In Virginia, emergency departments in critical access hospitals now use AI-powered tele-neurology to help rural teams make rapid stroke decisions. As Dr. Wendy Woolley described, patients can now be seen by a neurologist within minutes, using a system that supports triage and tele-rounding in-house.[6] The AI-assisted platform supports video-based assessments, camera control, and documentation integration. More importantly, it allows hospitals to retain patients when intervention isn't needed, preserving regional capacity and reducing unnecessary transfers. The lesson? AI works best when it supports *judgment*, not just speed. And it can extend the clinical reach of specialists to where they're needed most.

Financial Pillar: Cera employs AI-driven tools to predict and prevent hospitalizations among older and vulnerable individuals. Their technology analyzes patient data collected during home visits to identify health deterioration, enabling early interventions. This approach has reportedly reduced hospitalizations by up to 70 percent and patient falls by 20 percent, leading to significant cost savings for the UK's National Health Service (NHS). By shifting care from hospitals to patients' homes, Cera's model alleviates pressure on hospital resources and offers a cost-effective solution for long-term care, benefiting those with limited financial means. The company's services are largely funded through contracts with the NHS and local authorities, ensuring that patients incur minimal to no out-of-pocket expenses. This integration of AI into home healthcare exemplifies how technology can be leveraged to improve access to healthcare services **(Financial Barrier)**.[7,8,9]

---

[6] Woolley W. Interview with Matt S. Tele-neurology, rural care, and digital access. January 22, 2025. Accessed June 25, 2025.

[7] TechCrunch. UK in-home healthcare provider Cera raises $150M to scale its AI platform. January 12, 2025. Accessed April 12, 2025. https://techcrunch.com/2025/01/12/uk-in-home-healthcare-provider-cera-raises-150m-to-expand-its-ai-platform/. TechCrunch.

[8] The Times. Healthcare startup Cera wins unicorn status after raising $150m. January 13, 2025. Accessed April 12, 2025. https://www.thetimes.com/business-money/companies/article/healthcare-start-up-cera-wins-unicorn-status-after-raising-150m-x6dczkmwc.

[9] The Healthcare Technology Report. Cera secures $150M to expand AI-driven home healthcare. February 4, 2025. Accessed April 12, 2025. https://thehealthcaretechnologyreport.com/cera-secures-150m-to-expand-ai-driven-home-healthcare/.

In the United States, the integration of AI into health insurance processes has, in some cases, intensified the **Financial Barriers** to care, particularly in behavioral health and addiction treatment. Insurers have employed AI-driven systems to automate prior authorization and claims adjudication, often leading to denials that prioritize cost considerations over clinical necessity. United Healthcare utilized an AI tool known as ALERT, which was initially designed to identify patients at risk of suicide or substance use. This system was repurposed to flag what the company deemed "therapy overuse," targeting patients receiving frequent outpatient mental health services. The algorithm could trigger scrutiny when patients had therapy sessions twice a week for six weeks or more than 20 sessions in six months, leading to coverage denials without comprehensive clinical evaluation.[10]

Similarly, Cigna's subsidiary, EviCore, employs AI algorithms to assess the likelihood of claim approvals. These algorithms can automatically deny claims or refer them for further review, with the system's settings adjustable to meet internal "cost-saving" targets. Critics argue that such practices can result in the denial of medically necessary treatments, including those for behavioral health, based on rigid guidelines rather than individualized patient assessments.[11]

The American Medical Association (AMA) has expressed concern over the use of AI in prior authorization processes, noting that these systems can lead to systematic batch denials with little or no human oversight. Such practices may delay or disrupt patient access to essential medical treatments, particularly in behavioral health, where care often requires long-term, individualized interventions.[12] These developments highlight the need for greater transparency and oversight in the application of AI within health insurance, ensuring

---

[10] Elliott J. UnitedHealth used an algorithm to deny mental healthcare. It was probably illegal. *ProPublica*. Published March 5, 2024. Accessed April 16, 2025. https://www.propublica.org/article/unitedhealth-mental-health-care-denied-illegal-algorithm.

[11] Miller TC, Rucker P, Armstrong D. Inside the company helping America's biggest health insurers deny coverage for treatments. *ProPublica*. Published October 23, 2024. Accessed April 16, 2025. https://www.propublica.org/article/evicore-health-insurance-denials-cigna-unitedhealthcare-aetna-prior-authorizations.

[12] American Medical Association. Physicians concerned AI increases prior authorization denials. *American Medical Association*. Published March 4, 2024. Accessed April 16, 2025. https://www.ama-assn.org/press-center/press-releases/physicians-concerned-ai-increases-prior-authorization-denials.

that technological advancements serve to enhance, rather than hinder, patient care.

Cultural Pillar: In terms of cultural access, South Korea's AI-driven elder care robots such as Hyodol illustrate how cultural customization can support adoption. These robots are programmed to use Korean speech and traditional conversational patterns while delivering medication reminders, providing emotional companionship, and monitoring health behavior. The design reflects Korean social norms and respect for hierarchy, fostering trust among older adults who might otherwise reject impersonal digital tools.[13]

Digital Pillar: Meanwhile, as you may remember from Danny Gladden's interview, AI-powered telehealth systems introduced into Native Alaskan communities faced significant policy-induced limitations.[14] Even though remote consultations were technically feasible, Medicaid regulations required that patients physically present themselves at certified facilities to be eligible for telehealth reimbursement. This policy undermined the potential for digital access in geographically isolated regions and exposed the disconnect between digital tool deployment and structural regulation.

Trust/Knowledge Pillar: The final pillar is closely tied to data integrity and inclusive governance. In the United Kingdom, the NHS AI Lab has prioritized public engagement through initiatives like the 2022 Public Dialogue on Data Stewardship. This program brought together patients, clinicians, ethicists, and data scientists to deliberate on ethical data use, consent, and algorithmic fairness in AI development. Insights from these dialogues have informed the design and governance of AI tools, particularly in sensitive areas such as intensive care. By embedding public perspectives into AI governance and publishing transparency reports, the NHS has fostered greater trust among patients and providers regarding AI's role in clinical settings.[15]

---

[13] Lee OE, Yun JC. Perceptions and experiences of Korean American older adults with companion robots through long-term use: a comparative analysis of robot retention vs. return. *Front Public Health.* 2024;12:11609076. doi:10.3389/fpubh.2024.11609076. https://www.ncbi.nlm.nih.gov/pmc/articles/PMC11609076/.

[14] Gladden D. Interview with Matt S. Systemic gaps in access. March 7, 2025.

[15] Ipsos. *NHS AI Lab Public Dialogue on Data Stewardship.* NHS England and Sciencewise. Published November 2022. Accessed April 16, 2025. https://sciencewise.org.uk/wp-content/uploads/2022/11/22-033229-01-NHS-AI-Lab-Data-Stewardship-Dialogue-Report_0.pdf.

Conversely, several United States-based AI diagnostic tools have come under scrutiny for racial and gender bias. Many were trained predominantly on datasets from white, male patients, resulting in diagnostic inaccuracies for women and people of color. A 2024 investigation found that some FDA-cleared algorithms performed significantly worse in minority populations. Public backlash and coverage in both academic journals and media outlets have eroded trust and prompted calls for federal oversight.[16]

Where AI Is Already Working (and How to Build on It)

If the first chapter of AI in healthcare was written in the language of pilot projects, the second chapter is being written in scale. Not everywhere, but in enough places to prove that progress is possible. We know that AI can stumble when it's built without context. But what happens when it *is* context-aware? When it is paired with good policy, integrated into clinical workflows, and deployed in service of actual patient needs?

Remote Monitoring That Actually Reaches Patients

For anyone who has been confused about *remote patient monitoring* (RPM) for the last five chapters, here's the deal: it's a simple three-word idea. "Remote," means the patients are usually at home and not in the confines of a hospital. "Patient," well duh, we are helping folks here, mostly with chronic conditions. Lastly is "monitoring." That's different than just your fitness watch when you track things but don't actually have the information analyzed or even check it on a regular basis. (Yup, I said it!) When we talk about real "monitoring," it can include everything from a nurse checking a patient's weight using a scale that sends data to a daily dashboard, to a Holter monitor transmitting real-time information that identifies patient risk and alerts both the patient and provider to intervene. So … RPM: more than your watch, less than being confined to a hospital bed.

---

[16] Leo R. Health rounds: AI can have medical care biases too, study reveals. *Reuters.* Published April 9, 2025. Accessed April 16, 2025. https://www.reuters.com/business/healthcare-pharmaceuticals/health-rounds-ai-can-have-medical-care-biases-too-study-reveals-2025-04-09/.

> **Fact Check: REACH VET[17]**
>
> 1. REACH VET uses 61 clinical indicators from EHRs to identify veterans at highest statistical risk of suicide, hospitalization, or crisis.
>
> 2. The system can flag elevated risk before clinical symptoms or crisis behaviors emerge.
>
> 3. Early evaluations link the program to fewer psychiatric hospitalizations and higher follow-through on scheduled care.
>
> 4. Since its nationwide rollout in 2017, REACH VET has been implemented across all VA medical centers and healthcare systems.

Wearables are everywhere, but meaningful remote monitoring goes beyond step counts and sleep scores (see Diagram 6.3). For patients with chronic conditions, especially heart failure or diabetes, AI-powered alerts can catch deterioration before a crisis occurs. FDA-cleared tools like those in the Apple Watch and Fitbit now include ECG sensors capable of detecting atrial fibrillation.[18,19] But it's not just heart conditions, AI is changing diabetes management as well. Continuous glucose monitors—like Dexcom and FreeStyle Libre—now use AI algorithms to predict glucose fluctuations before they happen, helping diabetics avoid dangerous highs and lows.[20,21]

But the real win happens when infrastructure supports the tool. As Woolley noted, when digital literacy, staffing, and reimbursement

---

[17] U.S. Government Accountability Office. *VA Suicide Prevention: VA Should Assess the Use, Performance, and Effectiveness of Its Suicide Prevention Model.* GAO-22-105165. Published November 2021. Accessed June 26, 2025. https://www.gao.gov/assets/gao-22-105165.pdf.

[18] Apple Inc. ECG app and irregular heart rhythm notification available today on Apple Watch. Apple Newsroom. December 6, 2018. Accessed July 27, 2025. https://www.apple.com/newsroom/2018/12/ecg-app-and-irregular-heart-rhythm-notification-available-today-on-apple-watch/.

[19] United States Food and Drug Administration. Apple Watch Series 4 ECG app cleared by FDA. 2018. Accessed June 25, 2025. https://www.fda.gov/news-events/press-announcements/fda-clears-apple-watch-ecg-app.

[20] The Healthcare Technology Report. Cera secures $150M to expand AI-driven home healthcare. February 4, 2025. Accessed June 25, 2025. https://thehealthcaretechnologyreport.com/cera-secures-150m-to-expand-ai-driven-home-healthcare/.

[21] FreeStyle Libre. *Continuous Glucose Monitoring System.* Abbott Diabetes Care. Accessed June 25, 2025. https://www.freestyle.abbott/.

align, these systems reduce readmissions and empower patients.[22] Conversely, in Alaska and elsewhere, Medicaid policy restrictions that *require* physical presence, even for telehealth-eligible tools, can collapse the promise of remote AI entirely. The takeaway? AI-powered monitoring works, but only when the **Digital Pillar** is treated as a public utility, not a luxury.

AI-driven remote monitoring is still limited by one major factor: *Access.*

AI-driven remote monitoring is still limited by one major factor: *Access.* Many of the patients who could benefit most from this technology (low-income, rural, and aging populations) don't have the broadband access or digital literacy needed to use these tools effectively. If infrastructure gaps aren't addressed, AI in remote monitoring risks reinforcing existing health disparities rather than closing them.

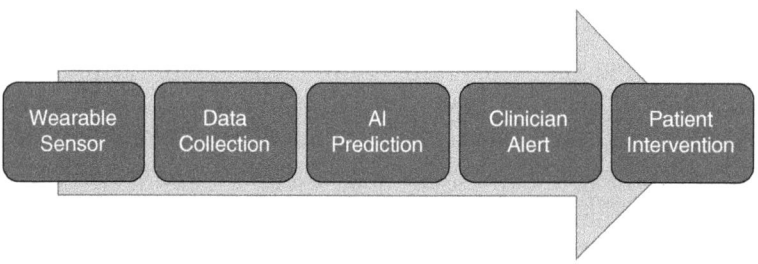

Diagram 6-3: How AI-Powered Remote Monitoring Tools Work.

How-To: Designing AI to Expand Access

Want AI to improve healthcare access, not just efficiency? Start here.

1. **Pair AI with Human Judgment:** Use algorithms to support, not replace, clinical decision-making. Tools work best when integrated into trusted workflows.

2. **Align Tools with Policy:** If you can't reimburse it, you can't scale it. Design around Medicaid, licensure, national health schemes, and billing constraints from the beginning.

---

[22] Woolley W. Interview with Matt S. Tele-neurology, rural care, and digital equity. January 22, 2025. Accessed June 25, 2025.

3. **Build for Real-World Constraints:** Design for bandwidth gaps, low digital literacy, and workforce shortages, not ideal conditions.

4. **Make Inclusion a Feature, Not an Afterthought:** Audit training data for demographic diversity. Involve patients and clinicians in product feedback and iteration.

5. **Measure What Matters:** Track reach, usability, and trust, not just technical performance or accuracy.

Stories from the Field: Trust/Knowledge Barrier Spotlight Nneka's Cupcake and the Butter of Racism[23]

For all its intelligence, AI is only as good as the data it's trained on, and that's where things get complicated. While AI has made significant strides in automating diagnostics and predicting disease risk, it has also revealed deep-seated biases that disproportionately affect marginalized communities.

Dr. Nneka Sederstrom doesn't mince words. "Racism is the butter of our healthcare system," she told me during one of the most unfiltered, honest interviews I've had for this book. "You can't just change the frosting or the sprinkles of a cupcake. You have to change the butter."

As a healthcare ethicist and leader at Hennepin Health in Minnesota, Nneka has spent her career unpacking the deep structural inequities that shape access to care. She sees how residential segregation, rooted in redlining that began in the southern states, has created healthcare deserts for people of color, both in inner cities and in rural areas. "Access," she said, "doesn't mean high-quality. It just means anything at all."

And that's where technology, including AI, often falls short. "Even if we put a pod out in rural Minnesota and screened someone via telemedicine," she said, "if there's no surgeon, no clinic, no real follow-up; what did we actually fix?" It's a haunting truth. AI may identify disease, but without the infrastructure to act on that diagnosis, the promise evaporates.

What hit hardest was her perspective on trust. "We've built healthcare for white people. And we expect communities of color to just plug into that and be grateful." She half laughed. "You can give someone a laptop and broadband, but if the only face on the other side of

[23] Sederstrom N. Interview with Matt S. Healthcare access and digital equity. Conducted virtually. March 31, 2025.

the screen is a white doctor with implicit bias, that isn't access (see Diagram 6.4). That's decoration."

Her cupcake analogy stuck with me. She explained how, for decades, healthcare systems have tried to improve inclusion by adding more diversity training, or funding digital tools in underserved areas. "That's changing the sprinkles," she said. "We've got to bake a new damn cupcake."

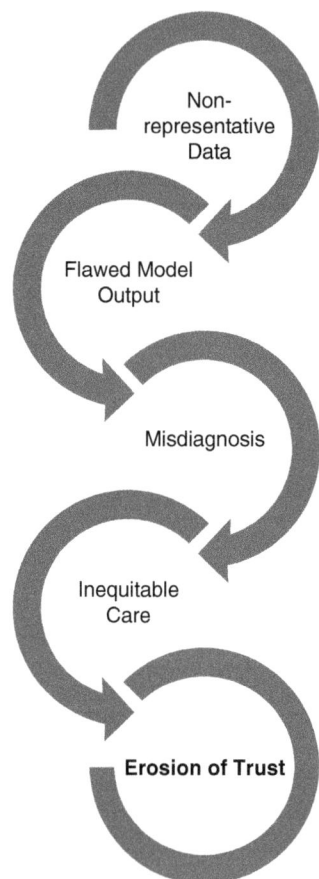

Diagram 6-4: The Bias Cascade in AI Diagnostic Tools.

We've got to bake a new damn cupcake.

An illustrative example of this issue emerged during the COVID-19 pandemic, when it was discovered that pulse oximeters, a staple in AI-driven remote monitoring, were far less accurate in detecting oxygen levels in patients with darker skin tones. This discrepancy led to delayed oxygen therapy for Black and Hispanic patients, contributing to higher COVID-19 mortality rates among these groups.[24]

Gender bias is a further concern. AI-driven cardiovascular risk models, long used to predict the likelihood of heart attacks, have historically underestimated risk in women, largely because most cardiovascular research has been conducted on male-dominated cohorts.[25] Addressing bias in AI requires more than just better programming: it demands a fundamental shift in how datasets are collected and used. If AI is to truly democratize healthcare, it must be designed for every patient, not just the ones who fit neatly into existing datasets.

Executive View: When AI Breaks the System Before It Fixes It[26]

For hospital executives today, the conversation about AI has already shifted. It's no longer "Should we use it?" It's, "What happens when it breaks the rest of our system?" Steve Bell, an experienced healthcare leader who has advised governments and global health systems, puts it bluntly: "You're not adopting a tool; you're agreeing to redesign how your hospital works." He challenges CEOs and Ministry of Health officials to confront AI's real cost—not just in capital but also in complexity. "The biggest risk isn't failure," he told me, "It's partial success that throws everything else out of balance."

He laid out three clear levels where AI needs to be understood systemically:

- **At the workforce level**, where generative tools like AI-powered Word and Excel can boost productivity but also shift the expectations and roles of non-clinical staff.

- **At the administrative level**, where AI alters hospital flow. Triage, scheduling, documentation, and capacity planning now hinge on predictive inputs.

---

[24] Sjoding MW, Dickson RP, Iwashyna TJ, Gay SE, Valley TS. Racial Bias in Pulse Oximetry Measurement. *N Engl J Med.* 2020;383(25):2477–2478. Accessed February 17, 2025. https://www.nejm.org/doi/full/10.1056/NEJMc2029240.
[25] Achtari M, Salihu A, Muller O, et al. Gender bias in AI's perception of cardiovascular risk. *J Med Internet Res.* 2024;26:e54242. Accessed February 17, 2025. https://www.ncbi.nlm.nih.gov/pmc/articles/PMC11538872/.
[26] Bell S. Interview by Matt S. Robotics, AI, and executive system design. Conducted virtually April 9, 2025.

- **At the clinical level,** where AI touches diagnoses, treatment decisions, and patient trust. These are areas where failure is unacceptable and transparency is non-negotiable.

> **Bring AI in first where it can save people, not where it replaces them.**

For leaders navigating this shift, Bell's advice is sharp: "Bring AI in first where it can save people, not where it replaces them." His strategy is to identify KPIs under stress and deploy AI in those high-friction zones; not to save money, but to save people from burnout and system failure. "Let AI arrive as the cavalry," he says, "not as the competition."

This executive lens reframes AI not as a shiny object or simple add-on, but as a full-system disruptor—an accelerator that can improve care, but also fracture workflows, staffing models, and trust if it's introduced too hastily. Bell's perspective is clear: Adopting AI means rethinking how a health system is built, from how we train our teams and manage information to how we fund operations and engage patients. In short, it touches every one of the five pillars. His call isn't to slow AI down. It's to get brutally honest about where it's most likely to succeed, and where, without systemic readiness, it might quietly erode the very trust **(Trust/Knowledge Pillar)** it's meant to build.

 How-To: Avoiding Common AI Pitfalls in Access Design

This one is for the builders (startups, in-house tech teams, and everyone in between):

1. **Building for the Well-Connected First:** If your solution only works with fast Internet, smartphones, or digital fluency, you're widening the gap.

2. **Prioritizing Precision over Participation:** An accurate model that no one uses is a failed tool. Focus on adoption, not just output.

3. **Ignoring Policy Realities:** Even the best algorithm stalls if Medicaid or the national health scheme won't reimburse its use. Design with regulators in mind.

4. **Skipping the End User:** AI built without patient and frontline input often misses cultural context, usability, and trust-building needs.

5. **Scaling a System That Was Already Broken:** Without addressing the disparities of the current system, scaling with AI often automates exclusion, not access.

Next Shift Quick Win: Use AI to Spot What's Missing

On your next shift, pick one patient flagged by AI. This could be a risk score, a fall risk, a sepsis alert, whatever.

Ask two things:

1. **Did anyone follow up?**

2. **Would this alert have helped last night?**

Write down what worked or didn't. Share it at huddle. That's it. You just made the algorithm better!

Rebuilding Trust (Not Just Technology)

Technology, no matter how advanced, cannot resolve healthcare's deepest inequities without trust **(Trust/Knowledge Pillar)**. While AI, telehealth, and wearables are transforming how care is delivered, their success ultimately hinges on whether patients and providers believe in them. Trust **(Trust/Knowledge Pillar)** is not an abstract ideal, it is built through usability, security, cultural alignment, and policies that protect people while enabling progress.

At the foundation of any trustworthy system lies the need for data protection. Patients must feel confident that their health information will not be misused. Cybersecurity breaches in healthcare have become increasingly common and costly. Over 640 incidents were reported in the United States in 2024 alone, exposing millions of patient records.[27] If you really want to scare yourself, check out the U.S. Department of Health and Human Services Office for Civil Rights Breach Portal.[26] There are breaches almost every day in the United States! As you can see from the portal, encryption, multi-factor authentication, and role-based access controls are not just best practices; they are ethical imperatives. In the meantime, I'm crossing my fingers that my data wasn't breached on that list somewhere.

Regulatory clarity is another cornerstone of trust. Without it, healthcare organizations hesitate to adopt AI and digital tools, fearing liability, reimbursement confusion, or conflicts with outdated laws. The World Health Organization (WHO) and national bodies like the

---

[27] U.S. Department of Health and Human Services. *Breach Portal: Notice to the Secretary of HHS Breach of Unsecured Protected Health Information*. Published 2024. Accessed April 5, 2025. https://ocrportal.hhs.gov/ocr/breach/breach_report.jsf.

FDA have begun to issue frameworks for AI governance, but challenges persist, especially in harmonizing global standards.[28]

Equally important is digital health literacy, not just for patients, but for providers and caregivers. Even the most promising AI or telemedicine solution will fail if users cannot navigate it or do not see its value. Studies show that digital health tools are underutilized and poorly understood among older adults, particularly those over 70, females, and those reporting relying on a helper.[29] Closing this gap requires more than technology; it demands human-centered design, community outreach, and culturally aligned education efforts.

Stories from the Field: Trust/Knowledge Barrier Spotlight[30]

In the early 2000s, Danny Gladden LCSW was working in HIV services in St. Louis when he received a call that would never leave him. A young man from a rural Missouri town had died just days after being admitted to an urban hospital. He was only in his early 20s.

"He went to the ER three times," Danny recalled. "Each time, they told him he had a bad cold or pneumonia. No one asked the right questions." On the fourth visit, this time to a hospital in St. Louis, a provider finally ran an HIV test. His viral load was in the hundreds of thousands. His CD4 count had dropped below 100. He had full-blown AIDS. The delay in diagnosis meant that by the time anyone named the problem, it was already too late.

"His mom called me," Danny said. "I was assigned to be his case manager, but she said, 'You'll get to the hospital before I do. Just say you're a friend.' He died before she arrived."

The young man's story wasn't an anomaly; it was a case study in how stigma, location, and systemic silence can work together to kill. He had grown up in a rural county where getting tested wasn't just inconvenient, it was dangerous. The local ER providers either didn't know to ask, didn't feel equipped to manage the results, or simply didn't see him as someone who could be at risk.

---

[28] World Health Organization. *Ethics and Governance of Artificial Intelligence for Health: Who Guidance*. Published June 28, 2021. Accessed April 5, 2025. https://www.who.int/publications/i/item/9789240029200.

[29] Czaja SJ, Boot WR, Charness N, Rogers WA, Sharit J. Improving digital health technology use in older adults: a mixed methods approach. *J Am Geriatr Soc.* 2021;69(11):3144–3152. doi:10.1111/jgs.18935. https://agsjournals.online library.wiley.com/doi/epdf/10.1111/jgs.18935/.

[30] Gladden D. Interview with Matt S. HIV misdiagnosis, rural stigma, and systemic gaps in access. March 7, 2025.

"He was gay and closeted," Danny explained. "He didn't want to be tested in his own town. He didn't want anyone to know. So, he kept showing up with symptoms and kept getting sent home." This young man's death underscores what happens when stigma, geographic isolation, and provider blind spots converge **(Cultural Barrier)**. AI isn't going to catch what no one asks for. Predictive analytics won't help when whole populations are invisible in the data, or when the human on the other end of the system refuses to see them.

Built to Last

For digital health to succeed, several structural elements must be in place. This is especially true for AI-powered tools. Each element maps to a pillar of access and must be built into the system, not added after the launch:

- **Digital Health Literacy Must Be Designed In, Not Taught Later:** Trust starts with usability. Tools must feel intuitive and human, not foreign, confusing, or condescending. That means designing for digital literacy from the start, not just for patients but also for caregivers and providers. The most advanced remote monitoring system is useless if it takes a 20-minute tutorial to send a blood pressure reading. A conversational AI fails if it leaves a Spanish-speaking grandmother confused or excluded. This is digital access, not just infrastructure. Fluency is key.

- **Cultural Relevance Must Outweigh Technical Novelty:** Trust also relies on cultural alignment. Tools must reflect the languages, values, and lived experiences of the communities they serve. This isn't just about translation, it's about meaning. A postpartum depression app built in Silicon Valley won't resonate in Nairobi unless it understands how mothers *there* experience care, stigma, and community. Cultural alignment isn't a nice-to-have. It's what makes technology usable in the first place.

- **Data Privacy and Security Must Be Non-Negotiable:** Without trust, digital health collapses. And in today's landscape of cyberattacks and health data breaches, trust begins with security. Patients will not use platforms they fear could expose their most sensitive information. Privacy, encryption, and access controls must be core design features, not compliance check boxes or end-stage patches. If people stop trusting the system, the system fails.

- **Regulatory Clarity Must Replace Guesswork:** Ambiguity kills adoption. Providers need to know when and how they can safely use AI tools. Uncertainty around liability, licensure, or billing makes even the best-designed technology unusable in practice. A nurse practitioner in rural Georgia cannot guess whether a remote glucose check counts toward metrics. A surgeon in Brazil cannot depend on an AI tool if no one can say whether it holds up in documentation or court. Technology can only scale when the rules are clear.

Trust in digital tools is not earned through branding or buzzwords. It is earned through usability, accountability, and respect. Without that foundation, no algorithm can overcome the structural failures of the system it's built within.

Infrastructure and Governance: Who Builds the Tools and for Whom?

The reality on the ground is stark: hospitals in rural and underserved areas are closing at alarming rates, clinicians are stretched beyond capacity, and broadband access remains uneven. For AI and digital health to be meaningful, it must be reliably reachable.

That means expanding broadband in ways that genuinely reach rural and tribal communities—not just in policy, but also in practice. Federal and state investments like the FCC's Rural Digital Opportunity Fund aim to close the gap, but implementation remains patchy and slow. Broadband is not just about access to video visits; it's the connective tissue for everything from diagnostics to remote monitoring. Bridging this digital divide is foundational to every other pillar of digital health.

We must also support the people who deliver care. Expanding loan forgiveness and incentive programs can draw clinicians to underserved areas, but we also need to acknowledge reality. Hybrid models that pair AI decision support with human clinicians may be essential to meeting demand. This is not about replacing the workforce, it's about reinforcing it and addressing any **Physical** and **Trust/Knowledge Barriers**.

Infrastructure also includes facilities. We need long-term, structural investment in rural healthcare systems: keeping critical access hospitals open, supporting community clinics, and ensuring that care is possible where people live. Public-private partnerships will be vital to sustaining services like maternity care, mental health, and emergency response. Without financial infrastructure, digital tools with new AI are just window dressing.[31]

But none of this works without trust **(Trust/Knowledge Pillar)**. Technology, no matter how advanced, cannot succeed if patients and pro-

> **None of this works without trust.**

viders do not believe in it. We often talk about trust as if it were an abstract ideal, a feeling to be inspired. In reality, trust is built the way bridges are: with structure, integrity, and careful attention to where the weight falls.

Trust, then, is not a feature, it is the foundation. Without it, access fails, inclusion stalls, and innovation collapses under the weight of its own good intentions. With it, we unlock the full potential of intelligent, connected care.

Governance and Inclusion[32]

While much of the conversation around AI and digital health focuses on capability, far less attention is paid to governance and inclusivity. As we strive for interoperable systems, we must ask: Who builds these tools, who decides how they are used, and most critically, who is left behind?

This last question was at the heart of a conversation with Dr. Brenda Ayers, a physician and equity advocate at Nuvance Health, who has spent years working at the intersection of data,

> **AI is only as good as the data you feed it, and the data we feed it is already biased.**

care, and community. "AI is only as good as the data you feed it," she explained, "and the data we feed it is already biased." Algorithms trained on historic data reflect historic inequities, from misdiagnoses to gaps in access and treatment. Without intervention, these tools do not fix healthcare disparities; they encode and accelerate them.

Dr. Ayer's solution is not just better data science, but better inclusion. "Representation," she said. "We build these tools without the people who will be most affected. Without women. Without Black and Brown communities. Without rural voices. And then we act surprised when the tools fail them." She also pointed to the limitations of policy alone: "Even if a tool works, it's meaningless if no one pays for

---

[31] Data Privacy Haiku: Whispers in the cloud, records float past firewalls thin, trust lost, hard to heal.
[32] Ayers B. Interview with Matt S. Health equity, data bias, and governance in AI. Conducted virtually. March 26, 2025.

it. Payers often block access to innovation, especially for marginalized groups. Access must be intentional. It doesn't happen by accident." Her most vivid metaphor? "AI right now is like watching a toddler with a loaded gun, and we're all tied to the chair."

Can AI Fulfill Its Promise? Only If We Start at the Margins

AI in healthcare is not an inevitability. It is a tool, and like all tools, its impact will depend on how and where we wield it. If we center AI development on already privileged systems and try to "scale out," we risk reinforcing the very problems we claim to solve. As Dr. Nneka Sederstrom reminds us, "You can't fix a racist cupcake by changing the frosting." It is time to bake a new recipe.

We need to stop sprinkling equity on top of broken systems. Inclusivity must be the foundation, not the garnish. That means designing for the margins from day one: in training datasets, in pilot studies, in broadband infrastructure, in workforce models, and in regulation. This is not just ethically urgent; it is practically essential.

AI will only close the gap if we:

- Build inclusive training data that reflects the real demographics of our communities
- Expand broadband and digital infrastructure in rural and tribal areas
- Develop sustainable, hybrid, human-AI care models, especially in underserved zones
- Prioritize cultural alignment and digital literacy in implementation
- Create regulatory pathways that reward tools improving real-world access

Stories like Danny's remind us what is at stake: a rural HIV patient misdiagnosed three times before anyone asked the right question, too late. AI might have surfaced his condition earlier, but only if the infrastructure existed to act on it. Intelligence without action is no cure.

Healthcare AI can either widen the chasm between privilege and systemic vulnerability or help bridge it. This is our inflection point. Will we retrofit inclusion into broken systems or build something better from the start? Will AI be another burden, or finally, intelligence without borders?

Startup Builder's Box: Building SAFE AI

If you're building digital tools for borderless care, this is where design meets reality. Use this builder's box as a high-pressure check against digital vaporware and false promises. SAFE AI is not about complexity; it's about credibility.

PILLAR	DESIGN QUESTIONS TO ASK	BUILD ACTIONS TO TAKE
Scalable	Will this work in a community clinic with 5 Mbps Internet and no tech support?	Pilot in a low-resource setting first. Make offline modes and fallback flows standard.
Actionable	Does this system make the next decision *clearer*, not murkier, for both the clinician and patient?	Require user testing with frontline clinicians and lay caregivers before releasing decision logic.
Fair	Who's invisible in the training data— rural teens, older adults, single dads, non-English speakers?	Run a representativeness audit. Add counterfactual test cases before go-live.
Embedded	Will this tool *fit into* the day, or *take over* the day?	Embed in one click within EHR workflows and patient portals. Hide or eliminate redundant fields.

Your build isn't ready if your users can't:

■ Explain the tool's recommendation to a confused patient.

■ Use it without retyping the same info three times.

■ Get help from a human when it breaks.

■ Name one group who would be harmed if it fails.

If you can't meet all four of these, it's not SAFE yet.

Leaders' End-of-Chapter Action Checklist: Chapter 6: "Intelligence Without Borders"

LEADER	HIGH-IMPACT ACTION TO STRENGTHEN TRUST AND KNOWLEDGE ACCESS IN AI-ENABLED CARE
❑ Board Director	Approve an annual SAFE AI audit and require executive attestation that no AI-enabled tools used in care delivery fail the Scalable, Actionable, Fair, or Embedded test
❑ Chief Executive Officer	Launch a "Plain Language AI" campaign to translate every patient-facing algorithmic tool into a fifth-grade reading level; require review by community advisors before public use
❑ Chief Information Officer	Post real-time uptimes, failure alerts, and fallback mode status for every clinical AI service (e.g., triage bots, scheduling predictions, risk scores) on an internal dashboard
❑ Chief Health Information Officer	Mandate that every AI-generated recommendation shown in the EHR includes source logic, last training data update, and a "Why this? Why now?" (doesn't need to be a popup, but needs to be available)
❑ VP Clinical Operations	Pilot two culturally localized AI models for high-volume workflows (e.g., maternal risk alerts, fall risk scoring) in different languages or communities; compare outcomes at 90 days
❑ VP Nursing & Patient Education	Integrate a "Digital Fluency Rounding Tool" into bedside practice; use it to screen patients for comfort with AI-enabled tools and route to live support or alternatives as needed
❑ VP Data & Analytics	Complete a structured fairness audit of all deployed clinical AI and machine-learning models; publicly share results, gaps, and improvement plans with affected care teams
❑ Telehealth Program Manager	Add a "trust moment" to every remote care onboarding script: include how AI tools are used, how data is protected, and when a human is in the loop
❑ Patient Experience Manager	Run a monthly AI mystery shopper program with real patients to test usability, transparency, and bias; report anonymized scores and stories to the Trust Council
❑ Community Health Worker Supervisor	Launch a "Know Your Algorithms" mobile popup series in community centers; offer hands-on demos of local AI tools and gather feedback on fears, myths, and misunderstandings
❑ Director of Snacks & Morale	Use your AI of choice to generate ten haikus about data privacy. Print them on biodegradable cupcake wrappers. Distribute responsibly (… *there might be one in this chapter's citations*)

Bridging Technology and Humanity to Create Sustainable Borderless Healthcare

On a gray January morning, Monique G, a 29-year-old substitute teacher in rural Howard County Indiana, woke with a sharp ache low in her belly; 60 miles of snow-covered highway stood between her and the closest urgent care. While her children finished breakfast, she opened Community Health Network's "Virtual Visit Now" button, chose the next slot that appeared, and was chatting with an NP before the school bus left the driveway. Ten minutes later the practitioner suspected early appendicitis, pushed an alert to the system hospital in Kokomo, and sent Monique directly to imaging. She was on an operating table within two hours and home again before the roads had fully thawed. Monique still teases that the app spared her both an ambulance bill and a night of worry, and she keeps it on her home screen because of the ease and engagement.[1]

Monique's story is common inside Community Health Network since the system replaced several disconnected scheduling tools with a single DexCare orchestration layer in early 2024. Capacity has

---

[1] DexCare. *Webinar Featuring Community Health Network: How to Tackle Capacity Issues and Treat More Patients*. Published April 2024. Accessed May 5, 2025. https://dexcare.com/resources/virtual-on-demand-best-practices-balance-capacity-during-peaks-in-demand/.

Diagram 7-1: Five Pillars Snapshot.

surged, with leaders reporting a 300 percent increase in the number of visits the virtual care team can handle each day, alongside a 46 percent jump in retail clinic traffic and a patient-return rate touching 50 percent within the same year.[2] Satisfaction mirrors the growth; two-thirds of all digital encounters come from repeat users and the mean Net Promoter Score sits comfortably above 99, nearly triple the industry baseline.[3] With scores this high, I wonder if it can also fold your laundry and make your morning coffee? (Next feature request: Drive my kids to soccer practice!)

The result feels less like technology and more like hospitality. Dedicated advanced practice providers, many working from home

---

[2] Raths D. More DexCare health system customers become investors. *Healthcare Innovation*. Published June 13, 2024. Accessed May 5, 2025. https://www.hcinn ovationgroup.com/clinical-it/digital-health-innovation/ news/55088527/more-dexcare-health-system-customers-become- investors.

[3] DexCare. *Webinar Featuring Community Health Network: How to Tackle Capacity Issues and Treat More Patients*. Published April 2024. Accessed May 5, 2025. https://dexcare.com/resources/virtual-on-demand-best- practices-balance-capacity-during-peaks-in-demand/.

offices, can step in for extra shifts during viral surges. A chat queue keeps patients informed while they wait, while routing algorithms steer first timers to brick and mortar care when video is not enough. For Monique, the intangible benefit is trust **(Trust/Knowledge Pillar)**. For Community Health Network, it's a data-backed demonstration that a thoughtfully designed digital front door can widen access, lift loyalty, and still pay the bills. This hopeful note offers the right doorway into this chapter before we explore the grittier street medicine landscape of Dr. David Lehmann's practice.

In healthcare, particularly when engaging marginalized populations, the concept of trust transcends mere patient-provider interactions (see Diagram 7.1); it forms the foundation upon which healthcare access is fundamentally built. Dr. David Lehmann's experience in street medicine vividly illustrates how trust **(Trust/Knowledge Pillar)** can act as a transformative bridge to healthcare delivery, particularly for populations that experience continuous exclusion from traditional medical services.

Stories from the Field: Trust/Knowledge Barrier Spotlight[4]

In April 2018, Dr. David Lehmann, a hospitalist at SUNY Upstate University in Syracuse, NY, was deeply frustrated by repetitive and superficial emergency department visits from a patient experiencing homelessness. He recognized that conventional healthcare was failing due to a profound lack of trust **(Trust/Knowledge Barrier)** and genuine engagement. "It really pissed me off," he recounted. This emotional response encapsulates both the anger toward systemic failure and the courage to pursue innovative, patient-centered solutions.

Partnering with the community organization, "In My Father's Kitchen," Lehmann directly addresses **Physical** and **Financial Barriers**, bringing essential medical services—including antibiotics, wound care, and even micro-dosing Suboxone treatments for opioid dependency—straight to street corners and other nontraditional settings. While the tangible medical services are critical, the underlying trust **(Trust/Knowledge Pillar)** forged through consistency and empathetic presence is paramount. The model recognizes that building trust requires sustained, longitudinal engagement. Patients

---

[4] Lehmann D. Interview with Matt S. Street medicine and healthcare access for the unhoused in Syracuse, NY. Virtual interview. March 20, 2025.

on the street often experience compounded mistrust due to systemic neglect, frequent stigmatization, and repeated experiences of care that disregard their individual needs. By embedding healthcare provisions within an already-trusted community outreach, Lehmann has been able to foster deeper relationships, effectively reducing barriers to healthcare access that conventional approaches often inadvertently reinforce.

Supporting Lehmann's approach, Hallie Buddendeck, a medical student deeply involved in the street medicine initiatives, emphasizes that trust **(Trust/Knowledge Pillar)** and relationships are built through ongoing, consistent interactions that acknowledge and respect patients' circumstances and autonomy.[5] According to Buddendeck, meaningful healthcare for individuals experiencing homelessness transcends medical treatments. It requires a genuine, sustained commitment to understanding their lived experiences, creating a space where patients feel seen, respected, and heard.

The rapid digital transformation of healthcare holds immense promise for creating borderless healthcare systems capable of delivering balanced and accessible care across geographic, cultural, and economic divides. Innovations such as telemedicine, AI-driven diagnostics, and remote patient monitoring signal a transformative shift in healthcare delivery. Yet, despite this technological potential, true success lies not in the technology alone but in thoughtfully addressing the complex human elements integral to healthcare: trust, empathy, cultural intelligence, digital literacy, and ongoing education.

Healthcare, at its core, remains profoundly human. Technological advancements, however revolutionary, are still tools—the means to the ultimate goal of improving human health and well-being. This chapter examines the essential human dimensions needed to ensure that digital healthcare remains not only technically sophisticated but also empathetically effective and culturally inclusive. By exploring real-world experiences from diverse healthcare environments—from street medicine programs in Syracuse, NY to rural telehealth initiatives for transgender populations and targeted interventions for veterans—this chapter highlights the necessity of placing humanity at the heart of digital health innovation.

---

[5] Buddendeck H. Interview with Matt S. Medical education and street medicine exposure among upstate NY trainees. Virtual interview. March 14, 2025.

> **Fact Check: Digital Empathy Can Travel**
>
> 1. In primary care visits, the mean Consultation and Relational Empathy score was 31.3 by telemedicine versus 33.8 in person. On a 50-point scale, this is less than a 5 percent variance.[6]
>
> 2. Among 451 remote radiation patients, 97.6 percent described provider communication as good to very good.[7]
>
> 3. Most responding remote radiation patients (87.8%) either preferred telehealth or expressed no preference for in-person vs. fully remote visits.[8]

Components of Trust

Trust does not arrive all at once; it accumulates like layers of lacquer that must each dry before the next is applied. The first layer is transparency of information. When patients can open their charts, read every clinician note, and challenge inaccuracies, they gain a sense of partnership rather than passive receipt of care. The Open Notes experience shows that such clear sight lines nurture confidence while also nudging professionals to write in plainer language and with greater care.[9]

That confidence deepens when digital encounters still feel human. Empathy and human connection need not fade across fiber-optic cables. For example, nurses who deliberately cultivate "digital empathy" report higher patient satisfaction and stronger therapeutic alliances.[10] They pause for micro-moments of silence, mirror facial expressions on

---

[6] Subramanya V, Spychalski J, Coats S, et al. Empathetic communication in telemedicine: a pilot study. *PRiMER*. 2024;8:36. doi:10.22454/PRiMER.2024.644242.

[7] Cuaron JJ, McBride S, Chino F, et al. Patient safety and satisfaction with fully remote management of radiation oncology care. *JAMA Netw Open*. 2024;7(6):e2416570. doi:10.1001/jamanetworkopen.2024.16570.

[8] Cuaron JJ, McBride S, Chino F, et al. Patient safety and satisfaction with fully remote management of radiation oncology care. *JAMA Netw Open*. 2024;7(6):e2416570. doi:10.1001/jamanetworkopen.2024.16570.

[9] Erlingsdóttir G, Petersson L, Jonnergård K. A theoretical twist on the transparency of Open Notes: qualitative analysis of health care professionals' free-text answers. *J Med Internet Res*. 2019;21(9):e14347.

[10] Abou Hashish EA. Compassion through technology: digital empathy concept analysis and implications in nursing. *Digital Health*. 2025;11:20552076251326221. Accessed May 4, 2025. https://pmc.ncbi.nlm.nih.gov/articles/PMC11907611/.

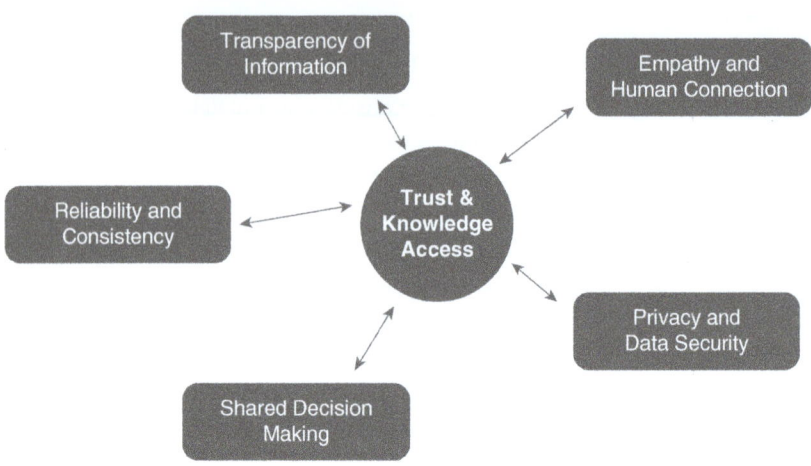

Diagram 7-2: Trust and Knowledge Access Components.

video, and use story-framing to convey understanding. These small acts signal that a real person, not an algorithmic avatar, is listening.

Consistency then seals the bond. Patients learn what to expect when every visit, message, and hand-off follows reliable patterns. The continuity tells them the system is steady, not unstable. A meta-analysis of 47 studies links such reliability and consistency to better subjective health outcomes, underscoring that trust **(Trust/Knowledge Pillar)** flourishes when routines are predictable and promises are kept.[11]

Yet even a consistent, empathic service can crumble if data are exposed. Privacy and data security safeguard the bond; breaches erode loyalty faster than any apology can repair. Cybersecurity scholars argue that zero-trust architectures and shared ownership of security practices among clinicians, administrators, and technologists are now prerequisites for credible digital care.[12]

Finally, genuine, shared decision-making moves trust from a nice idea to a true partnership in action. Systematic reviews show that when clinicians explicitly integrate patient values into choices—using option grids, plain-language probabilities, and signed care plans—patients

---

[11] Birkhäuer J, Gaab J, Kossowsky J, et al. Trust in the health care professional and health outcome: a meta-analysis. *PLoS One*. 2017;12(2):e0170988. Accessed May 4, 2025. https://pmc.ncbi.nlm.nih.gov/articles/PMC5295692/.

[12] Clarke M, Martin K. Managing cybersecurity risk in healthcare settings. *Healthc Manage Forum*. 2023;37(1):17–20. Accessed May 4, 2025. https://pmc.ncbi.nlm.nih.gov/articles/PMC10725101/.

adhere more tightly and report higher well-being.[13] Woven together, these five strands (transparent information, digital empathy, reliable consistency, uncompromising privacy, and collaborative decision making) form a braid strong enough to carry the weight of remote robotics and AI-enabled medicine into settings where trust has often been in short supply (see Diagram 7.2).

Human Connection

Digital health has transformed healthcare delivery, but its true power lies in its ability to foster genuine empathy and warmth in virtual interactions. Digital empathy involves leveraging technology to establish meaningful, compassionate care relationships, essential for effective virtual care. Telehealth has notably expanded access, particularly for mental healthcare in rural and underserved areas, directly addressing barriers posed by all the barriers—**Physical**, **Financial**, **Cultural**, **Digital**, and **Trust/Knowledge**. As digital methods become mainstream, maintaining empathy traditionally associated with in-person care remains essential.

Stories from the Field: Trust/Knowledge Pillar Spotlight

The case of Hannah, a 45-year-old mother in East Austin, exemplifies digital empathy effectively tackling the **Trust/Knowledge** and **Cultural Pillars**. Hannah, an African American woman embedded in a tight-knit family environment, faced multiple comorbidities, including severe obesity, diabetes, and hypertension. Previous negative healthcare experiences created mistrust and hesitation. Initially apprehensive, Hannah found comfort in her provider's genuine interest and active listening (fortunately that provider was me!). "Doctor," Hannah expressed, visibly moved, "nobody's actually listened to me before. They just see a fat lady who can't control herself." (Full disclosure: The appointment was perilously close to 2 p.m.; my glucose and my patience both dip at 1:59 p.m. Saved by the clock.)

---

[13] Tringale M, Stephen G, Boylan AM, Heneghan C. Integrating patient values and preferences in healthcare: a systematic review of qualitative evidence. *BMJ Open.* 2022;12(11):e067268. Accessed May 4, 2025. https://pubmed.ncbi.nlm.nih.gov/36400731/.

My approach included an immersive understanding of Hannah's environment—her living space, dietary habits, family dynamics, and social challenges. These factors are often invisible in traditional clinical care. Back in the mid 2000s, telehealth was not as mainstream as it is today, and my Medicare house call practice allowed me into her home to grasp her medical symptoms alongside underlying social determinants affecting Hannah's overall health. Empathy here was profoundly human, extending beyond clinical assessments to encompass environmental, familial, and emotional contexts. When I visited patients in their "real lives," it was often very easy to understand why they struggled with their diabetes, why they couldn't get to other providers' appointments, or why it was impossible for them to quit smoking. Realistic management strategies were needed that aligned with Hannah's daily life, thereby building the essential **Trust/ Knowledge Pillar**. While I was able to build this trust one visit at a time in her home, we need to do the same thing, even in a digital or virtual context.

Internationally, digital empathy addresses **Physical** and **Cultural Barriers** significantly. In the rural Kirehe District of Rwanda, a community health worker (CHW)-led, home-based telemedicine program for post-cesarean wound follow-up was viewed as highly acceptable by every mother and community health worker interviewed.[14] Women highlighted the CHWs' empathetic home visits and the trusted relationship they shared as key reasons they felt engaged and supported, while the virtual model removed long travel burdens and strengthened trust **(Trust/Knowledge Pillar)**, even in the face of physical isolation **(Physical Pillar)**.

Recent literature on AI highlights its emerging role in empathetic patient care, particularly addressing the **Trust/Knowledge Pillar**. AI-driven chatbots and virtual assistants have demonstrated the capacity to provide empathetic responses, with studies finding that AI-generated interactions can match or exceed human empathy in certain contexts, positively influencing patient engagement and adherence. Furthermore, AI systems designed to recognize cultural nuances

---

[14] Bikorimana L, Estrada EH, Niyigena A, et al. Acceptability of telemedicine for early surgical site infection diagnosis after cesarean delivery in rural Rwanda: a qualitative study. *Maternal Health, Neonatal Perinatal.* 2025;11:3. doi:10.1186/ s40748-024-00200-9.

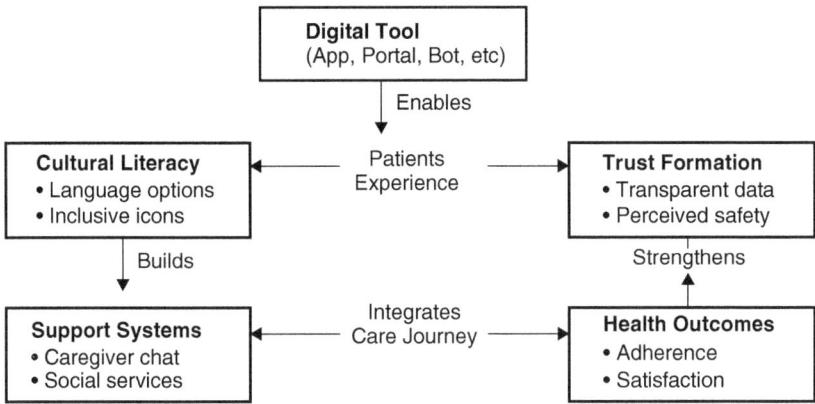

Digital Empathy in Action—Questions to Ask Before Launching a Tool:
1. Have we validated language and imagery with the target population?
2. Can the tool connect users to live human support when needed?
3. How will we measure sustained trust over time?
4. Do data-sharing practices align with community values?

Diagram 7-3: Empathy in Virtual Care: A Framework.

can effectively tailor communications to diverse patient populations, fostering trust and cultural sensitivity.[15]

> **Dr. Matt gets tired, Dr. Matt gets hangry, and Dr. Matt can have a bad day. You know who doesn't have a bad day? AI.**

I am often asked about the ability of AI to be empathetic, and I'm sure it can be a heck of a lot more empathetic than Dr. Matt (that's me) on lots of occasions! Dr. Matt gets tired, Dr. Matt gets "hangry," and Dr. Matt can have a bad day. You know who doesn't have a bad day? AI. Consistency is the magic it can provide to patients. Not all doctors have the bandwidth to be empathetic with every patient and in every situation, because they are human too.

AI integration in healthcare holds potential for reducing financial concerns **(Financial Barrier)** by enhancing efficiency in administrative tasks and preliminary patient assessments, thus broadening accessibility to quality care. Despite these promising advancements, researchers emphasize that AI-generated empathy should supplement

---

[15] Ayers JW, Poliak A, Dredze M, et al. Comparing physician and artificial intelligence chatbot responses to patient questions posted to a public social media forum. *JAMA Intern Med.* 2023;183(6):589–596. doi:10.1001/jamainternmed.2023.1838.

rather than replace human interactions, as AI currently lacks the genuine emotional understanding and authenticity inherent in human empathy (see Diagram 7.3). Ethical considerations such as data privacy, the authenticity of interactions, and the risk of technological dependence must be addressed to ensure AI enhances empathetic healthcare delivery.[16]

Digital Empathy How-To: Conducting Tele-Visits

■ **Frame Face-to-Face:** Position the camera so the lens is at your eye level. Keep a plain backdrop and light your face from the front; this arrangement restores eye-contact cues and raises patient trust scores in video visits.[17]

■ **Pause and Signal:** After each patient statement, allow a brief two-second silence, then offer a micro-acknowledgment such as a nod, slight smile, or "I hear you." The added beat absorbs audio delay and your cue reassures the patient that you are listening.[18,19]

■ **Name the Feeling:** Mirror the emotion you hear with a concise phrase ("that sounds exhausting," "you seem relieved") and confirm understanding. Explicitly labeling emotion in a digital setting achieves empathy scores comparable with in-room visits and softens perceived power distance.[20]

■ **Narrate Your Clicks:** When you look away to review records or type notes, say what you are doing ("I am opening your lab results now")

---

[16] Morrow E, Zidaru T, Ross F, et al. Artificial intelligence technologies and compassion in healthcare: a systematic scoping review. *Front Psychol.* 2023;13:971044. doi:10.3389/fpsyg.2022.971044.

[17] Tewksbury C, Deleener ME, Dumon KR, Williams NN. Practical considerations of developing and conducting a successful telehealth practice in response to COVID. *Nutr Clin Pract.* 2021;36(4):769–774. doi:10.1002/ncp.10742.

[18] Lee JF, Schieltz KM, Suess AN, et al. Guidelines for developing telehealth services and troubleshooting problems with telehealth technology when coaching parents to conduct functional analyses and functional communication training in their homes. *Behav Anal Pract.* 2014;8(2):190–200. Accessed May 7, 2025. https://pmc.ncbi.nlm.nih.gov/articles/PMC5048262/.

[19] Guetterman TC, Sakakibara R, Baireddy S, et al. Medical students' experiences and outcomes using a virtual human simulation to improve communication skills: mixed methods study. *J Med Internet Res.* 2019;21(11):e15459. Accessed May 7, 2025. https://www.jmir.org/2019/11/e15459/.

[20] Subramanya V, Spychalski J, Coats S, et al. Empathetic communication in telemedicine: a pilot study. *PRiMER.* 2024;8:36. Accessed May 7, 2025. https://journals.stfm.org/primer/2024/subramanya-2023-0125/.

and return your gaze to the lens. This transparent commentary maintains connection while necessary documentation continues.

▪ **Close with Teach Back:** End by asking the patient to restate the plan in their own words. Correct gently and share the written summary on-screen. Systematic teach back improves recall and adherence across literacy levels.[21]

Successful digital empathy requires intentionality: understanding patient lives beyond clinical symptoms, addressing cultural nuances, and fostering real-time connections that build trust, just like I did in person with Hannah. While digital tools offer unprecedented access, they must be intentionally designed and implemented with empathy as a core principle, ensuring technology enhances rather than diminishes the patient-provider relationship. Digital empathy emerges as an essential framework within digital health, facilitating deep human connections that transcend **Physical Barriers**. Whether in rural America or global healthcare settings, embedding digital empathy into telehealth practices ensures that care is not just accessible, but meaningful, responsive, and profoundly human.

Cultural Intelligence and Inclusivity

In an era marked by increasing globalization and cultural interconnectedness, healthcare delivery is challenged not merely by **Physical Barriers** but significantly by complex **Cultural Barriers** and linguistic differences. Achieving genuine healthcare access requires more than just physical availability of services; it necessitates meaningful engagement, cultural sensitivity, and intentional inclusivity that permeate through every interaction between patients and providers. This underscores the necessity for cultivating cultural intelligence, the capabilities to function effectively across diverse cultural contexts and within healthcare practices and systems.

Dr. Lehmann's street medicine programs in Syracuse offer a powerful example of addressing cultural intelligence through direct community engagement. Their work highlights not only the **Physical Pillar**

---

[21] Brach C, ed. *Health Literacy Universal Precautions Toolkit.* 3rd ed. Agency for Healthcare Research and Quality; 2023. Tool 5: Use the Teach-Back Method. Accessed May 7, 2025. https://www.ahrq.gov/sites/default/files/publications2/files/health-literacy-universal-precautions-toolkit-3rd-edition.pdf.

(bringing care directly to where patients are), but also emphasizes the **Cultural Pillar** (cultural sensitivity), the **Financial Pillar** (financial accessibility), and the **Trust/Knowledge Pillar** (trust building). The team approaches patients with a profound understanding of the local cultural nuances and socioeconomic circumstances, recognizing that without this understanding, patients are unlikely to engage meaningfully with their own healthcare. For example, addressing a predominantly unhoused population involves more than just treating medical issues; it requires navigating complex social and psychological dimensions such as substance abuse, chronic mental health conditions, and deeply embedded mistrust of traditional healthcare institutions.[22]

Similarly, insights shared by Dr. Nneka Sederstrom of Hennepin Healthcare provide a deeper view into how cultural intelligence **(Cultural Barrier)** directly intersects with issues of racial bias and structural racism. She asserts that healthcare disparities in Minnesota, and across the United States, cannot be fully addressed without confronting the legacy of racism that shapes patients' lived experiences, especially in marginalized and rural communities. Sederstrom vividly explains how healthcare institutions, deeply embedded with systemic biases, continue to disadvantage racial minorities through inadequate representation among providers, culturally inappropriate healthcare policies, and ineffective communication that fails to resonate with diverse patient populations.[23]

Scholarly research consistently underscores the need for healthcare professionals to possess cultural competence and intelligence, which has been shown to significantly improve patient engagement and clinical outcomes. A framework by Betancourt et al. emphasizes that culturally competent healthcare systems recognize the importance of culture in healthcare delivery and adapt services to meet patients' social, cultural, and linguistic needs.[24] Saha et al. further expands on this by asserting that culturally competent care not only reduces health disparities but also enhances overall healthcare quality and patient satisfaction.[25]

---

[22] Lehmann D. Street Medicine Discussion. Virtual interview. Interviewed by Matt S. March 20, 2025.

[23] Sederstrom N. Discussion on Healthcare Access. Virtual interview. Interviewed by Matt S. March 31, 2025.

[24] Betancourt JR, Green AR, Carrillo JE, Ananeh-Firempong O II. Defining cultural competence: a practical framework for addressing racial/ethnic disparities in health and healthcare. *Public Health Rep.* 2003;118(4):293–302. Accessed April 21, 2025. https://journals.sagepub.com/doi/abs/10.1093/phr/118.4.293.

[25] Saha S, Beach MC, Cooper LA. Patient-centeredness, cultural competence, and healthcare quality. *J Natl Med Assoc.* 2008;100(11):1275–1285. Accessed April 21, 2025. https://www.ncbi.nlm.nih.gov/pmc/articles/PMC2824588/.

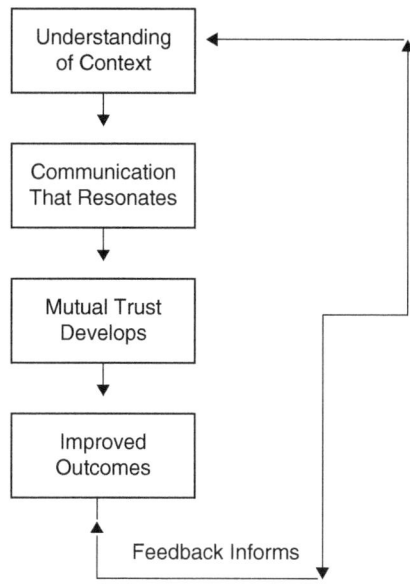

Diagram 7-4: Culturally Competent Care Cycle.

Stories from the Field: Cultural Pillar Spotlight

Singapore's HealthHub app lets every resident toggle instantly among English, Mandarin, Malay, and Tamil. The same four languages appear in the companion Healthy 365 wellness app and on printed Healthier SG care plans.[26,27] By embedding language choice in the default user interface rather than providing it as a separate translation layer, the platform removes both cultural and literacy friction (see Diagram 7.4). Ministry of Health data shows that more than 700,000 residents enrolled in Healthier SG within the first nine months, with engagement highest in districts where Malay or Tamil is the primary home language.[28] Behind the scenes, Synapxe, Singapore's national health-tech agency, maintains a single clinical terminology set but surfaces it through four parallel glossaries vetted by

---

[26] HealthHub Support. *Multi Language FAQ.* Accessed May 7, 2025. `https://support.healthhub.sg/hc/en-us/sections/26050053835417-Multi-Language`.
[27] HealthHub. *Healthy 365 App Guides.* Accessed May 7, 2025. `https://www.healthhub.sg/programmes/healthyliving`.
[28] Ministry of Health Singapore. *Enhancing Preventive Health and Aged Care.* March 6, 2024. Accessed May 7, 2025. `https://www.moh.gov.sg/newsroom/enhancing-preventive-health-and-aged-care`.

community linguists. This workflow keeps vocabulary aligned across lab results, prescription instructions, and chatbot answers while respecting cultural nuance. The lesson for cultural intelligence is clear: bake language into code and governance from day one rather than bolting it on later.

Despite these advancements, there remains significant skepticism regarding the authenticity and effectiveness of culturally inclusive approaches. As Dr. Brenda Ayers of Nuvance Health points out, inherent biases in AI algorithms, a rapidly growing component of digital healthcare, could exacerbate rather than alleviate disparities if the data underpinning these systems remains culturally homogenous or biased. Ayers underscores the critical necessity of inclusive governance in AI development, ensuring diverse representation among decision-makers and direct community involvement to mitigate potential bias. Without this inclusivity at the foundation **(Cultural Pillar)**, digital health risks amplifying existing imbalances.[29]

Stories from the Field: Digital Barrier Spotlight

Diane Francis didn't plan to spend over two decades at Johnson & Johnson (J&J).[30] She was supposed to gain experience, then head to U Penn for a PhD in public health. But the intersection of science, public health, and market strategy held her fast. She quickly discovered that healthcare wasn't just about treating disease. It was about creating demand for interventions that people didn't even know they needed and confronting the systemic blind spots baked into those decisions.

Her first major wake-up call came during a lung cancer screening initiative. As she combed through outcome data, something wasn't adding up. Traditional analysis focused on comparing population averages; disparities were framed in terms of how marginalized groups fared relative to white patients. But Diane saw a deeper flaw. "We focus on the average," she said, "but the average doesn't represent the population." The true insights, she realized, were hidden at the edges—among the best and worst outcomes. Could understanding those extremes uncover not just gaps, but missed opportunities?

---

[29] Ayers B. AI Bias Discussion. Virtual interview. Interviewed by Matt S. March 26, 2025.
[30] Francis D. Interview with Sarah Matt. Virtual interview. Interviewed by Matt S. February 19, 2025.

That hypothesis came into sharper focus during the CABANA Trial, which evaluated atrial fibrillation outcomes for ablation therapy. When Diane reviewed the paper, the main narrative barely mentioned subgroup differences. But buried in a supplementary table, she found what should have been a headline: while Black men had a lower recorded prevalence of atrial fibrillation, they responded *better* to ablation than to medical therapy. Meanwhile, Asian women showed significantly higher complication rates. A follow-up, real-world Medicare study confirmed it. These weren't statistical footnotes, they were clinical truths with life-or-death implications. And yet, they'd been overlooked.

This realization—that data disparity isn't just a measure of inequity but a roadmap to better care—set Diane on a new path. At Edwards Lifesciences, she turned her attention to structural heart disease, where traditional eligibility criteria routinely excluded the very patients who might benefit most. Mary, whom Diane came to call "Mrs. Avocado," became the living embodiment of this journey.

Stories from the Field: Digital Pillar Spotlight

At 72, Mary's heart was failing. She had end-stage tricuspid regurgitation, and her heart had swollen to the size of an avocado. Her case had been dismissed as untreatable. Hospice was called. She didn't qualify for intervention, not by the rigid clinical metrics used at the time. But Diane saw her differently. Mary had been screened early, a crucial but often dismissed first step. That meant there was a window.

Diane and her team didn't just question the rules, they redefined them. Mary was offered a tricuspid valve replacement, an option once considered too risky or too "futile." But for Mary, the decision was clear: "What do I have to lose that I haven't already lost?"

Four years later, she was walking three miles a day. Her life had not only been extended, but it had also been reclaimed. She became an outspoken advocate for others with tricuspid disease, helping shed light on what many still call "the forgotten valve." Mary's recovery wasn't just a clinical win. It was proof that the system's exclusions, often framed as cost-effectiveness or clinical conservatism, could be deadly. Patients were willing to take risks if it meant a better life. The question wasn't just about safety; it was about whose version of safety we were privileging.

Mary's recovery spurred Diane to rethink access entirely. After COVID-19, she revisited trials once written off as marginal, and hidden disparities emerged. Black patients appeared only at crisis stage; women were scarce in procedural registries; care pathways favored theoretical efficiency over lived reality. The system had not merely withheld treatment, it had failed to notice those most in need (see Diagram 7.5). Ignore the outliers and we erase the very people innovation could save. Mary's avocado-sized heart now stands as proof that inclusive care hinges less on bigger datasets and more on the courage to ask who is missing and why.

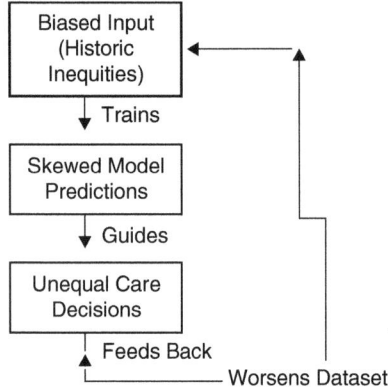

Diagram 7-5: Bias Feedback Loop in AI Healthcare.

Fact Check[31]

1. Almost 25 percent of consumers say they would switch physicians if virtual visits were not offered.

2. Sixty-four percent of U.S. consumers rate virtual appointments more convenient than in-person care.

3. In 2024, nearly all consumers who have had a virtual visit (94%) expressed a willingness to have another one, up from 80 percent in 2020.

---

[31] Deloitte Center for Health Solutions. *The Growing Disconnect Between Virtual Health Availability and Consumer Demand.* Published October 16, 2024. Accessed May 5, 2025. https://www2.deloitte.com/us/en/insights/industry/health-care/virtual-health-consumer-demand-and-availability.html.

Ultimately, healthcare access demands an intentional shift toward genuine cultural intelligence and inclusivity. Healthcare providers must move beyond superficial accommodations to cultivate an authentic understanding and appreciation of diverse patient backgrounds. Programs such as Syracuse's street medicine initiatives and Diane Francis's patient-driven healthcare analytics, as well as the literature, illustrate that culturally intelligent healthcare practices not only reduce imbalance but also build trust and improve health outcomes across communities. To achieve sustained change, healthcare systems must continuously integrate cultural intelligence training, inclusive digital tool development, and inclusive policy reforms as core strategies for bridging healthcare disparities.

 Next Shift Quick Wins:

1. **Transparency Flashcard:** Read your plan paragraph aloud; press Sign; show the patient the portal timestamp before closing the visit.

2. **Pause, Smile, Mirror:** Tape these three words under your webcam and glance at them before every virtual encounter to reinforce digital empathy.

3. **Privacy Promise Opener:** Start each video call with, "Your data travels in an encrypted tunnel just like online banking; if you prefer voice only, tell me now," then wait three seconds for response.

Training and Education: Preparing Providers and Patients for Digital Health Success

In today's rapidly evolving healthcare landscape, the transformation toward digital solutions such as telemedicine, AI, and remote patient monitoring promises unprecedented improvements in healthcare access. However, this digital shift also introduces considerable challenges, particularly around digital literacy and effective integration into daily medical practice. Successfully navigating this transition requires structured, intentional training and education programs. These programs must not only address technical competencies but also emphasize cultural sensitivity, accessibility, and sustained engagement—core elements aligned with the five pillars of healthcare access.

Stories from the Field: Trust/Knowledge Barrier Spotlight[32]

Brooks grew up in San Antonio but moved to a remote college town in East Texas, where pursuing gender-affirming care meant driving five hours round-trip to Houston just to fill a prescription. The clinic misfiled their initial paperwork, and because testosterone is a controlled substance, the pharmacy wouldn't budge. Brooks made that trip multiple times: facing stares, suspicion, and outright hostility.

"Luckily, I had a car," Brooks said. "But most people don't. If I hadn't been in college, if I had a job or kids, I probably would have given up."

Their journey didn't start with a supportive healthcare team. The first therapist they saw pressured them to freeze their eggs, start hormones, and book surgery, all within a few sessions. Brooks wasn't ready. When they hesitated, the therapist dropped them. That betrayal of trust and agency almost derailed their entire transition. In time, Brooks found a better path. They now live with a supportive community, but their story exposes how easily people can be pushed out of care by systems that assume compliance, speed, and digital fluency. Access isn't just about geography. It's about respect, relevance, and resilience.

Stories from the Field: Digital Barrier Spotlight[33]

For Lucky, the barriers started early. Growing up in Wayne County, New York, they always knew something didn't fit, but it took years to discover the language and community that gave shape to their identity. Once they began seeking care as a nonbinary adult, Lucky ran into a wall of digital friction. "I would literally bribe friends to sit with me and help me schedule an appointment," they told me. "The systems were so rigid and overwhelming, and I didn't trust anyone I was calling."

When they finally moved to Rochester, NY, a door opened. Through a little-known trans health program, Lucky was assigned a navigator who could manage both the bureaucratic noise and the emotional burden. The care team wasn't just affirming; they were accessible by text, not just through a labyrinth of portals and phone trees. "I finally felt like someone was advocating for me *with* the system, not just within it," Lucky said.

---

[32] Brooks. Interview with Matt S. Gender-affirming care, systemic gatekeeping, and digital navigation barriers. February 26, 2025.
[33] Lucky. Interview with Matt S. Nonbinary identity, digital health friction, and community-centered care. March 5, 2025.

Still, the residue of exclusion lingers. At the local pharmacy, Lucky would often send their spouse to pick up prescriptions just to avoid being misgendered or questioned. "I got weird looks," they said. "I'd hand over my name, and they'd glance at me like, *that doesn't match.* It was humiliating." Lucky's story underscores a core truth: a system that is designed around binary assumptions and digital self-navigation will never be accessible to everyone. AI won't fix that. People will.

Digital literacy, or the capacity to effectively use digital tools, emerges consistently as a significant barrier to healthcare access **(Digital Barrier)**. Interviews with Brooks and Lucky provide compelling personal insights, emphasizing that digital literacy extends beyond simple technical capabilities. For individuals in rural communities or those under-going sensitive health transitions, such as transgender individuals, **Digital Barriers** can compound feelings of isolation and mistrust in healthcare systems. Brooks described encountering significant barriers related to digital interfaces, logistical complexity, and lack of empathetic support from healthcare providers when accessing transgender health services remotely. For Brooks, digital literacy wasn't just about using technology, it was about navigating a healthcare system that often felt unwelcoming or unprepared to meet their specific needs. Training providers in both technical use and cultural competencies could significantly reduce these barriers by fostering more empathetic and knowledgeable interactions with patients, thereby enhancing patient trust and adherence to digital healthcare recommendations.

Stories from the Field: Digital Pillar Spotlight

A pilot at the Veterans Affairs (VA) Medical Center in Asheville, North Carolina, drops participants into *Rotate,* a virtual-reality (VR) app where they must save a medieval village from a fire-breathing dragon while physical therapists track cervical and shoulder movement in real time. Veterans play at home, and clinicians watch a live data stream scaling the difficulty without a single commute. Early results are promising, with program leaders reporting stronger adherence because the sessions feel like play rather than work.[34] One module even turns lower extremity drills into a pinball game guided by foot taps, making "leg day" an on-screen adventure.

---

[34] Hennick C. Telehealth and virtual reality expand into mainstream care at the VA. *HealthTech Magazine.* Published October 30, 2023. Accessed May 12, 2025. https://healthtechmagazine.net/article/2023/10/telehealth-and-virtual reality-expand-mainstream-care-va.

The pilot is part of VA Immersive, a nationwide effort that now counts more than 1,200 VR headsets spread across 160 medical centers and clinics, touching every U.S. state and Puerto Rico.[35] Veterans who have experienced the program overwhelmingly want more VR and rate the experience an 8 out of 10 on average.[36]

Momentum accelerated in April 2023, when the Office of Healthcare Innovation and Learning announced it was expanding remote rehabilitation to six sites: Richmond, Asheville, Los Angeles, Des Moines, Orlando, and Eastern Colorado.[37] The bundle pairs a head-mounted display with limb sensors and a library of clinically grounded activities for orthopedics, neurorehabilitation, and pain management. The hardware ships directly to veterans so care can run inside the living room.

The same platform helps with pain and behavioral health. Results have shown incredible breakthroughs, such as patients lifting an injured shoulder higher than they have in 20 years while placing a virtual baby bird back into its nest. The immersion distracts them from fear of movement and reduces dependence on medication. There have also been parallel gains in anxiety and depression management, reinforcing that VR is becoming a versatile clinical staple rather than a niche novelty.

What began as a dragon-slaying game now shows clinicians a blueprint for remote, data-rich rehabilitation that scales social connection alongside range of motion.

The Learning Curve How-To: Building Digital Confidence at Every Level

Why It Matters: Digital transformation stalls when patients cannot navigate portals, clinicians feel under-prepared for virtual care, or organizations lack a safe place to rehearse new workflows. Closing those gaps requires a tiered learning ecosystem that grows literacy, skills, and institutional muscle together (see Diagram 7.6).

[35] Bailey AL. VA expands virtual reality for Veteran rehabilitation. *VA News.* Published April 3, 2023. Accessed May 12, 2025. https://news.va.gov/117227/va-expands-virtual-reality-rehabilitation/.
[36] Hennick C. Telehealth and virtual reality expand into mainstream care at the VA. *HealthTech Magazine.* Published October 30, 2023. Accessed May 12, 2025. https://healthtechmagazine.net/article/2023/10/telehealth-and-virtual-reality-expand-mainstream-care-va.
[37] Bailey AL. VA expands virtual reality for Veteran rehabilitation. *VA News.* Published April 3, 2023. Accessed May 12, 2025. https://news.va.gov/117227/va-expands-virtual-reality-rehabilitation/.

STEP	CORE SKILL	WHY IT MATTERS	COMMON SLIP-UP
1	Power and battery check	A session ends the moment a device powers down	Laptop left unplugged; battery gives the farewell beep during consent
2	Secure Wi-Fi or approved VPN	Protected traffic prevents data drift and credential theft	Clinician joins on guest network; encounters video jitters, then drops
3	Camera framing and lighting	Clear eye contact nurtures empathy and diagnostic confidence	Back-lighting silhouettes the speaker; patient sees a mysterious outline
4	Microphone test and ambient noise control	Intelligible audio halves visit length and error rate	Barking dog at clinician side; caregiver hears only woofs
5	Browser and link verification	Correct URL shields against phishing and malware	User pastes meeting link into search bar; arrives at an ad farm
6	Click precision, single versus double	Proper clicking reduces accidental form resubmissions that corrupt data	Veteran double-clicks submit; portal files duplicate requisitions
7	Portal navigation and lab retrieval	Patients gain timely insight without playing phone tag	Results buried behind Show More button; frustration rises
8	File upload and screenshot ethics	Secure images accelerate wound assessment and diminish email ping-pong	Caregiver texts incision photo; metadata lost, privacy risk grows
9	Password stewardship with two-factor authentication	Unique passphrases and codes seal the front door	All devices share "Password123"; audit triggers emergency reset

(continues)

(Continued)

STEP	CORE SKILL	WHY IT MATTERS	COMMON SLIP-UP
10	Controlled update staging	Testing on a sandbox device averts clinic-wide outages	Beta roll-out at 8 a.m.; entire waiting room freezes by 8:01
11	Data-flow awareness and basic API vocabulary	Interoperability underpins longitudinal care and analytics	Medtech calls API a new sushi roll; team laughs, data remains siloed
12	Peer mentoring and micro-teaching	Teaching spreads competence faster than any help desk	Mentor forgets to un-mute; class still learns (and someone brings tacos)

Diagram 7-6: Digital-Literacy Ladder.

Patient Digital-Literacy Ladders

A 2025 *Journal of Medical Internet Research* study validated a 10-item Digital Health Literacy scale for adults older than 65. The scores correlated directly with portal logins and tele-consult completion.[38] Programs that move the needle pair skills coaching with social motivation. When patient navigators phoned patients the day before video visits to walk through portal log-in, camera checks, and backup options, visit attendance climbed from 82.8 percent to 91.6 percent over 12 weeks. There was also 9 percent drop in no-shows, while producing a positive return on investment.[39] A larger follow-up across 17 primary care sites in the same system later confirmed a 3.7 percent relative reduction in non-attendance when navigators were present.[40]

---

[38] Kim J, Park E, Lee H. Measuring digital health literacy in older adults. *J Med Internet Res.* 2025;27:e65492. Accessed May 7, 2025. https://www.jmir.org/2025/1/e65492.

[39] Mechanic OJ, Lee EM, Sheehan HM, et al. Evaluation of telehealth visit attendance after implementation of a patient navigator program. *JAMA Network Open.* 2022;5(12):e2245615. doi:10.1001/jamanetworkopen.2022.45615.

[40] Chen K, Katranji K, Bailey K, et al. Effect of a telehealth navigator program on video visit scheduling and completion in primary care. *J Prim Care Community Health.* 2024;15:21501319231225997. doi:10.1177/21501319231225997.

Key takeaways for practice:

- Start with a short, validated screen; repeat every six months to gauge progress.
- Offer communal coaching spaces so learning happens with peers, not alone.
- Reward early wins, such as first secure message sent, to sustain momentum.

Clinician Micro-Credentialing

Frontline clinicians often receive only a single orientation slide deck before being thrown into virtual visits. Two emerging micro-credential models fill the gap. Angsana Health's Fundamentals of Telehealth program in Malaysia delivers asynchronous modules on regulation, device setup, and "webside manner," concluding with a live simulation graded by standardized patients.[41] In the United States, a one-hour equity-focused telehealth simulation raised communication scores and increased satisfaction by patients.[42] Micro-credentials travel with the clinician's professional profile, making competence visible to employers and payers.

Design principles:

- Keep total time under eight hours so the course fits into busy shifts.
- Combine bite-size theory with high-fidelity roleplay for muscle memory.
- Issue portable digital badges that link to demonstration artifacts.

Organizational Practice Sandboxes

Hospitals need a safe, rule-bounded environment to prototype new virtual-care features without full regulatory burden. Indonesia is piloting a health sandbox that lets startups test tele-malaria tools

---

[41] Angsana Health. *Fundamentals of Telehealth Microcredential*. Accessed May 7, 2025. https://www.angsanahealth.com/en/fot/.
[42] Viswanathan M, Cross S, Jones L. Telehealth equity and access communication skills pilot simulation. *Med Teach*. 2025;47(2):123–130. Accessed May 7, 2025. https://pubmed.ncbi.nlm.nih.gov/39761252/.

under real-world supervision; early feedback loops inform national policy before wide deployment.[43] It's possible the same sandbox model could cure "pilotitis" by replacing scattered small pilots with staged scale-ups governed by clear guardrails.

Action checklist for leaders:

- Carve out a regulatory light zone governed by ethics review and rapid audit.

- Invite cross-functional teams (IT, compliance, frontline staff, etc.) to co-design tests.

- Publish sandbox results in an open dashboard to build confidence and attract partners.

Combined, these patient, clinician, and organizational ladders create a self-reinforcing learning curve that shortens time to safe adoption and widens access for every demographic.

Fact Check: The Digital-Readiness Divide

1. Americans answered a median of five of nine items correctly on a national digital-knowledge test.

2. On the digital knowledge test, only 48 percent could pick out a two-factor-authentication example, underscoring the skills gap.[44]

3. U.S. smart-phone ownership is 96 percent among adults 18 to 29, and 61 percent in those 65 and up. This is a 35-point gap.[45]

---

[43] Siregar E, Putri D, Oktaviani S. Introducing a regulatory sandbox into the Indonesian health system. *J Med Internet Res.* 2023;25:e47706. Accessed May 7, 2025. https://www.jmir.org/2023/1/e47706/.

[44] Pew Research Center. *What Americans Know About AI, Cybersecurity and Big Tech.* Published August 17, 2023. Accessed May 5, 2025. https://www.pewresearch.org/internet/2023/08/17/what-americans-know-about-ai-cybersecurity-and-big-tech/.

[45] Pew Research Center. *Share of Those 65 and Older Who Are Tech Users Has Grown in the Past Decade.* Published January 13, 2022. Accessed May 5, 2025. https://www.pewresearch.org/short-reads/2022/01/13/share-of-those-65-and-older-who-are-tech-users-has-grown-in-the-past-decade/.

Champions and Advocates: Catalysts for Digital Adoption

Stories from the Field: Cultural Pillar Spotlight[46]

Dr. Kenneth Johnson knows what it means to fight, both on the battlefield and off. As a former military medic and now a healthcare innovator, Kenneth has spent decades witnessing how veterans fall through the cracks of a fractured system. His story begins at Fort Bragg, the largest military base in the world by population. Despite its size, Fort Bragg is what Kenneth calls "a sand-filled void," a place with few local resources for the men and women who return from war with physical and mental wounds.

Kenneth describes veterans flying across the country, sometimes even crossing international borders, to access care. Psychedelic therapy in Mexico. Rehabilitation programs in San Antonio. Hyperbaric oxygen treatments in Florida. "We put them on planes and send them off to get the help they need," he says, "but when they come home, there's nothing. No follow-up care. No local support."

Determined to change that, Kenneth is building what he calls "a hub for healing." His vision? A center **(Physical Pillar)** where veterans can access therapies like transcranial magnetic stimulation, physical rehabilitation, counseling, and even community activities that foster connection. He's creating a space that prioritizes health and wellness over the "drinking culture" often tied to veteran social clubs. "Veterans need a place where they can rebuild their lives—one that celebrates health, not just war stories over a bar," Kenneth says.

But he's not stopping there. Kenneth knows that a physical hub can only go so far. For veterans in remote areas, like Wyoming or Alaska, technology must fill the gap. Drawing on his military experience, where medics used encrypted telehealth systems to consult specialists halfway around the world, Kenneth plans to integrate telehealth, wearables, and patient-controlled digital records powered by AI into his nonprofit. "The technology has to work for them," he explains. "It needs to be simple, valuable, and built with veterans in mind."

Kenneth is also tackling another challenge: the fractured system of support for veterans. With hundreds of organizations operating in silos, each with its own resources and goals, Kenneth envisions

---

[46] Transcript discussion with Kenneth Johnson, January 21, 2025.

a hub-and-spoke model that unites these groups. "We need to stop working in isolation and start working together. Veterans deserve better."

For Kenneth, this mission is deeply personal. "Too many of the guys I served with are frustrated, texting me for help because the VA isn't meeting their needs," he says. "We can't just patch the system with Band-Aids. We need to build something new, something that lasts."

Dr. Johnson's experience underscores the necessity of specialized training programs for healthcare providers. Veterans often exhibit unique challenges related to technology engagement, shaped by military experiences and frequently co-occurring mental health disorders. Johnson highlighted the critical importance of culturally tailored provider training that incorporates an understanding of military culture, trauma-informed care approaches, and digital health proficiency tailored to the veteran community.[46] Culturally tailored educational interventions can significantly improve healthcare outcomes among veterans, particularly when combined with digitally integrated care models.

Local "champions" play pivotal roles in energizing digital adoption within healthcare settings. These influential individuals—often frontline providers—bridge **Cultural, Digital**, and **Trust/ Knowledge Pillars** by demonstrating the practical and clinical value of digital interventions. Physician advocates in Saitama Prefecture illustrate how clinician support can bridge Japan's regulatory and **Cultural Barriers** for AI-assisted triage. An emergency medicine team from the Saitama Medical Association, NEC, and the prefectural government developed an AI chatbot to guide residents through symptom checks and recommend ambulance use. During a six-week pilot in Spring 2019, the service averaged about 100 consultations daily and received positive feedback, leading to prefecture-level safety clearance and a full-scale launch.[47] Early evaluations showed its value for younger users who prefer texting and patients with hearing impairments, two groups often missed by telephone helplines. Pro tip: The bot never puts you on hold to pet the office Shiba Inu.

---

[47] NEC Corporation. Chat type emergency consultations, The future emergency medical service: AI supports humans. *NEC Insights*. Published May 21, 2021. Accessed May 11, 2025. https://www.nec.com/en/global/insights/article/2021050002/index.html.

Likewise, successful street medicine programs such as those led by Dr. David Lehmann illustrate how local champions can leverage trust **(Trust/Knowledge Pillar)** and show cultural competence **(Cultural Pilar)** to effectively integrate digital tools into non-traditional health-care settings, enhancing **Physical (Pillar)** and **Financial (Pillar)** access.[48]

Ultimately, education and training initiatives must fundamentally aim to build trust. Providers who are culturally competent and digitally fluent can meaningfully engage diverse patient populations, fostering trust and increasing adherence to healthcare interventions. Culturally sensitive communication training and digital literacy significantly improve patient-provider relationships, directly impacting clinical outcomes. Culturally competent training has been proven to not only improve clinical outcomes but also reduced disparities among diverse patient populations.[49] Similarly, Saha et al. emphasized that culturally competent and patient-centered communication training significantly enhances patient trust and satisfaction, particularly within historically under-served communities.[50]

As healthcare continues its trajectory toward digital transformation, a comprehensive approach to training and education remains indispensable. By addressing digital literacy barriers, tailoring training to specialized populations, and leveraging local champions, healthcare organizations can bridge the gap between digital technology and human-centric healthcare, fostering trust and promoting universal, effective healthcare delivery across diverse populations.[51]

---

[48] Lehmann D. Street Medicine Discussion. Virtual interview. Interviewed by Matt S. March 20, 2025.

[49] Betancourt JR, Corbett J, Bondaryk MR. Addressing disparities and achieving equity: cultural competence, ethics, and health-care transformation. *Chest.* 2016;149(1):143–148. doi:10.1378/chest.13-0634.

[50] Saha S, Beach MC, Cooper LA. *J Natl Med Assoc.* 2008;100(11):1275–1285.

[51] *If you are reading this at 2 a.m., please hydrate.*

Startup Builder's Box: Startup Moves That Hardwire Trust and Knowledge

STARTUP MOVE	NEXT SPRINT ACTION	TRUST-PROOF METRIC
Transparent by default	Release clinician notes or lab results the moment they are signed; run a 10-patient readability check; adjust wording to eighth-grade level (or less!)	≥90% of users open their note within 24 hours
Open APIs invite scrutiny	Stand up complete HL7 FHIR endpoints; demonstrate an end-to-end pull from a test EHR in an acceptable timeframe during demos	Integration time to first data pull ≤ x minutes
Teach before you treat	Embed six-minute micro-learning modules that cover app use, condition basics, and next steps; measure knowledge gain pre- and post-module	Mean knowledge score rise ≥ 10% after one session
Seal credibility early	Launch a SOC2 readiness gap review this quarter; assign one founder as control owner for every Trust Services criterion	Close all critical gaps and earn SOC2 report within nine months
Secure every click	Implement a zero-trust architecture with device discovery, micro-segmentation, and real-time risk scoring; publish uptime and incident metrics on your status page	Critical security incidents per year = 0

Use this box whenever your team decides what to build next; each line turns an abstract principle into a measurable test you can hit in one release cycle.

Leaders' End-of-Chapter Action Checklist: Chapter 7: "Bridging Technology and Humanity to Create Sustainable Borderless Healthcare"

LEADER	HIGH-IMPACT ACTION TO STRENGTHEN TRUST AND KNOWLEDGE ACCESS
❑ Board Director	Endorse a quarterly public Trust Index and tie 5% of executive bonus pool to its year-over-year rise
❑ Chief Executive Officer	Launch a cross-functional Trust Council within 30 days and charge it with a 12-month action plan covering transparency, empathy, consistency, privacy, and shared decision making
❑ Chief Information Officer	Post live telehealth uptime and latency metrics on the intranet home page and review weekly for outliers
❑ Chief Health Information Officer	Standardize after-visit summaries to include the clinician's plan paragraph verbatim; track monthly EHR compliance and nurse completion of a 60-minute digital empathy micro-credential
❑ VP Clinical Operations	Embed option-grid decision aids in the five most common virtual encounters and verify patient understanding with a two-question quiz
❑ VP Nursing & Patient Education	Assign a single care-coordinator-of-the-week for each clinic and display that contact on the patient portal every Monday
❑ VP Data & Analytics	Complete a bias and privacy audit of all predictive models that affect triage or scheduling and present findings to the Trust Council
❑ Telehealth Program Manager	Script and enforce a 30-second "privacy promise" greeting for virtual visits; sample 10 encounters per month to confirm delivery
❑ Patient Experience Manager	Install one-tap emoji feedback buttons on portals and kiosks; send a weekly wins-and-fixes digest to frontline teams
❑ Community Health Worker Supervisor	Run monthly pop-up digital cafés teaching portal logins; aim to raise activation rates and cut missed appointments by 10% in six months
❑ Director of Food and Morale	Buy cupcakes without systemic bias; share at next Trust Council

Economics of Borderless Care

Why Capital Matters for Access

I remember finally getting to schedule my appointment for the first COVID-19 vaccine in 2020. My family had been in close quarters with four little kids, a cat, and a dog for way too long. The idea of being able to safely see grandparents and friends was a huge draw. But getting access to the vaccine was a totally different ballgame. First, I couldn't get an appointment for months anywhere nearby. In fact, the system only allowed a patient to sign up within the next two weeks at most pharmacies in the United States. You might remember at the beginning of the vaccine roll-out there were no approved child vaccinations, so this was just going to be for me and my husband to start. We tried every location within an hour radius, and our primary care doctor couldn't even get access to the vaccines for use in their own office. As a medical professional who understands the system, has transportation, insurance, and broadband, I still struggled. One of my family members had a little summer cabin about an hour and a half away and we decided to look and see if we could get appointments nearby. Amazingly we were able to grab appointments the next weekend and jumped at the chance.

The whole family (including the cat and the dog) packed into the truck and headed south. We finally arrived down a dirt road to a little cabin in the woods, with no cell reception but reasonable Internet. There were no stores or gas stations nearby. The next neighbor was farther than the eye could see, and if you had an emergency, you'd pretty much be toast. The next day was our vaccination day. We left early just in case. It took over 30 minutes to get to the appointment. As I think now to that small community and how they must have obtained their healthcare on a regular basis—they couldn't get any care, or *anything* for that matter without at least a 30-minute drive. We passed three Dollar General stores on the way there, as these were the most abundant storefronts in the area. When we arrived at the pharmacy, we masked up and went in to get vaccinated in shifts so we could keep the kids' disease free in the car. A whole weekend trip, and this was just dose number one for the adults. We'd have to do the same thing to get our second vaccine, and we'd still have to wait to get the kids squared away as well.

The story of my family's vaccine journey illustrates the complex challenges of healthcare access during the pandemic and highlights critical issues—transportation barriers even for those with means, rural pharmacy and healthcare deserts, vaccine distribution imbalance, multiple-dose logistics challenges, and countless others (see Diagram 8.1).

If we were to tackle this problem of vaccine access in rural environments, there probably wouldn't be an easy button. We'd start with small pilots and see what works, iterate, and eventually scale. But no matter how compelling a pilot project may be, it quickly stalls without steady financial resources. In earlier chapters, you learned how telehealth or AI ventures in India, Kenya, and other regions risk failing when hardware maintenance or software upgrades are unfunded after the initial launch. Capital extends beyond startup costs; it includes the day-to-day and year-to-year budget that keeps server capacity robust, reimburses staff for telehealth sessions, and ensures connectivity in remote zones. Connecting to the pillars: Capital is pivotal for the **Financial Pillar**, aligning with "sustainable reimbursement" structures, but it also intersects with infrastructure requirements (e.g., equipping rural clinics with devices, ensuring stable electricity) and **Digital Pillar** buildouts (broadband capacity, software maintenance).

Diagram 8-1: Five Pillars Snapshot.

Economic Realities

No matter how sophisticated a digital health solution may be, financial stability remains a core determinant of whether it scales beyond the pilot phase. A robust healthcare system—one with clear regulations, broad stakeholder buy-in, and interoperable infrastructure—naturally draws in diverse financing streams. Governments can allocate budget lines for telehealth services; public and private insurers may reimburse virtual consultations; and public-private partnerships (PPPs) can co-fund emerging AI diagnostics or remote monitoring programs.

In contrast, a single tech platform dropped into a fragile system often struggles for funding because payers and investors view it as too risky. Japan's Medical DX Promotion Plan explicitly allocated funding for EHR modernization and telemedicine pilots; however, local hospitals still struggled with the high cost of ongoing AI development, showcasing the tension between initial policy backing and real-world budget constraints.

Affordability How-To: Five Practical Moves to Eliminate Sticker Shock

STEP	WHAT TO DO	WHY IT MATTERS
1. **Deliver a digital estimate before service.**	Text or portal-push a personalized out-of-pocket estimate 72 hours in advance for every scheduled encounter; provide links to lower-cost alternatives when available	Patients rank unexpected medical bills as their top financial fear;[1] early visibility increases adherence and lowers no-show rates
2. **Embed real-time pharmacy benefit checks.**	Prompt prescribers with onscreen alerts that show copay, prior authorization status, and same-class generics while they write the prescription	Cost transparency at the point of decision increases the switch to low-cost options and cuts primary non-adherence
3. **Offer zero-interest payment plans at checkout.**	Enroll any balance over $200 in an automated 6–12 month plan with no added fees; surface charity eligibility screens for self-pay patients	Structured plans reduce bad debt by up to 30% while preserving patient satisfaction[2]
4. **Publish a public affordability scorecard.**	Display median out-of-pocket spend per episode, surprise bill complaints, and charity approval rates on the organization website quarterly	Price transparency rules give hospitals a legal nudge; visible metrics create peer comparison pressure[3]
5. **Guide consumers to independent cost tools.**	Add a "compare local prices" button that deep links to FAIRHealthConsumer. com or a state all payer claims database	External tools complement hospital lists and help patients spot large regional price spreads[4]

---

[1] KFF. *Americans' Challenges with Health Care Costs.* Issue brief. Published March 10, 2024. Accessed May 19, 2025. https://www.kff.org/health-costs/issue-brief/americans-challenges-with-health-care-costs/.

[2] Solomon J, Lessard L. CarePayment program hospital outcomes: results from semi-structured interviews with hospital staff. *J Health Care Finance.* 2016;42(4). Accessed May 21, 2025. https://healthfinancejournal.com/index.php/johcf/article/view/66.

[3] Ibid.

[4] FAIR Health. *FH Consumer Classroom.* Accessed May 19, 2025. https://www.fairhealthconsumer.org/.

Next Shift Quick Wins: Quick Deployment Tips

- Start with the 10 highest volume shoppable services; expand once workflows stabilize.
- Use plain language and round numbers, for example, "about two hundred dollars," to aid comprehension.
- Track conversion from estimate viewed to visit completed as your leading indicator.

Systems Thinking, Scarce Dollars

Let's talk about the quiet power of money; not billion-dollar medtech IPOs, but the slower, structural flow of cash inside hospital walls. Because the way a health system earns, spends, and saves can dictate whether a patient sees a clinician tomorrow or waits weeks for an appointment that never opens up.

Most hospitals don't earn revenue from routine care. They earn it in operating rooms, through diagnostic imaging, or via reimbursed specialty services. Primary care and emergency departments often run at a loss. Innovation programs? They usually don't generate revenue at all. Which means margin (typically just 2%–4%) is not a rounding error; it's the oxygen line for strategy. When that line tightens, the path narrows. Even the most exciting digital health tool may get shelved unless it can clearly cut costs or drive reimbursable value within a 12-month window.

The decision to adopt new technology doesn't rest in one office. It moves through a friction-filled relay: clinicians who want better tools, IT

Margin is not a rounding error; it's the oxygen line for strategy.

teams who fear security risks, procurement departments chasing discounts, and CFOs scrutinizing anything that touches the balance sheet. Without clear alignment, even proven solutions can stall. And the tighter the margin, the more conservative the mindset. Hospitals prefer vendors with *scale* (large enough companies to offer multiple solutions and a track record of success with delivery systems), tools that work with existing workflows, and partners who understand that every dollar committed to new technology means one less dollar for direct patient care. But financial rigor does not have to mean stagnation. Some leaders use those constraints to push for smarter, faster transformations.

Stories from the Field: Financial Pillar Spotlight[5]

Dr. Al Villarin, CMIO of Nuvance Health in the NY and CT Hudson Valley, has walked this tightrope. As a practicing physician and informatics leader, Dr. Villarin wanted to improve documentation efficiency using AI-based scribe tools. But before he could pitch any pilot, he had to align the clinical benefit with his health system's operational and financial constraints.

Most CMIOs, he explained, report into the IT structure, where budget control often stops at infrastructure. Meanwhile, clinical and finance teams operate in parallel but not always in sync. "That's where things get complicated," he said. "You can have a great solution, but if it lives in the wrong part of the org chart, it never gets funding."

So, his team flipped the script. Instead of leading with a cool product, they led with a financial goal: cut documentation time by two minutes per visit. They translated those minutes into hard savings and productivity gains that could be tied back to staffing models and provider satisfaction metrics. Then, they aligned the pilot with broader institutional metrics: retention, throughput, and burnout reduction.

It worked. Not only did the pilot win approval, but it showed early promise on both the clinical and financial sides. "We didn't ask for innovation dollars," Al said. "We asked for a return. And we showed them where to find it." That mindset—earn your budget with measurable ROI—has since shaped Nuvance's broader digital transformation strategy. Clinical champions still play a role. But innovation must prove its value not just to patients or IT, but to the P&L.

Dr. Villarin's experience offers a roadmap for founders and innovators trying to gain a foothold inside complex systems. If your solution doesn't solve a budget-line problem, it doesn't matter how smart your AI is. "What's been most effective," he says, "is showing how we remove friction for the people who already feel the pressure—clinicians and the finance team alike." This is how innovation scales in the real world: not by bypassing the budget, but by becoming indispensable to it.

When Free Is the Only Price People Can Pay

In rural America, **Financial Barriers** aren't about high deductibles or network gaps, they're about *nothing*. No cash. No coverage. No car. In Appalachia, the reality of **Financial Barriers** is not abstract. It's daily,

---

[5] Matt S. Interview with Al Villarin, MD, Chief Medical Information Officer, Nuvance Health. May 30, 2025.

physical, and fatal. While national systems wrangle over value-based models and coverage tiers, one small nonprofit has built a parallel economy of care without any billing infrastructure at all. The Health Wagon doesn't just lower cost. It eliminates it. In doing so, it offers a radical challenge to the conventional economics of care, especially in places where the market has already walked away.

Stories from the Field: Financial Pillar Spotlight[6,7]

In the coalfields of southwest Virginia, where jobs have vanished, chronic illness is rampant, and insurance coverage is often a luxury, **Financial Barriers** to care remain the most significant barriers. The Health Wagon, a nonprofit mobile medical unit founded by Sister Bernadette Kenny in 1980, has spent over four decades proving that healthcare can be both free and sustainable when it's built on the **Trust/Knowledge Pillar**, community partnerships, and fiscal creativity. Serving some of the most economically distressed counties in the United States, the Health Wagon's work has become a case study in the **Financial Pillar**—not as a theoretical right, but as a daily, operational reality.

Every service the Health Wagon provides is completely free: primary care, medications, labs, dental and eye clinics, even specialty consultations. In a region where many residents fall into the Medicaid coverage gap or work in informal economies without benefits, eliminating cost is not just compassionate, it's essential. Patients have no copays, no premiums, and no deductibles. They are never asked for a credit card. And yet, the program remains viable. It survives not through billing cycles but through a braided funding stream of grants, public university partnerships, community donations, and national sponsors. In the early years, the organization operated on less than $2 million, delivering care to thousands at a per-capita cost that undercut almost any other care delivery model in the country.

Partnerships with institutions like the University of Virginia School of Medicine and Remote Area Medical (RAM) amplify both funding and service reach. The Health Wagon was one of the first rural programs in the United States to integrate telehealth services, allowing

---

[6] Gardner T, Gavaza P, Meade P, Adkins D. Delivering free healthcare to rural Central Appalachia population: the case of the Health Wagon. *Rural Remote Health.* 2012;12:2035. Accessed June 4, 2025. https://www.rrh.org.au/journal/article/2035.

[7] The Health Wagon – Mobile Clinics. Accessed June 4, 2025. https://thehealthwagon.org.

Health Wagon's Funding Flywheel

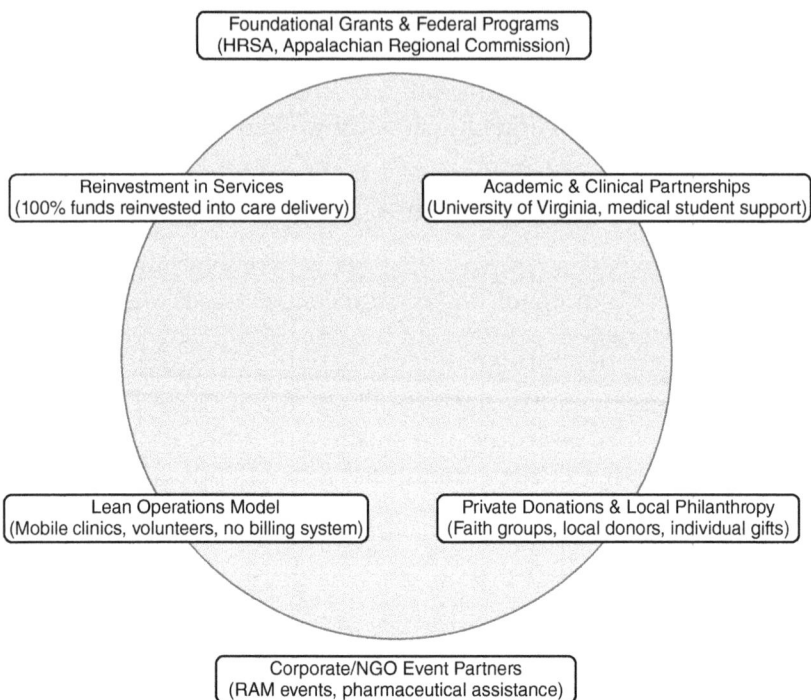

Foundational Grants & Federal Programs
(HRSA, Appalachian Regional Commission)

Reinvestment in Services
(100% funds reinvested into care delivery)

Academic & Clinical Partnerships
(University of Virginia, medical student support)

Lean Operations Model
(Mobile clinics, volunteers, no billing system)

Private Donations & Local Philanthropy
(Faith groups, local donors, individual gifts)

Corporate/NGO Event Partners
(RAM events, pharmaceutical assistance)

Diagram 8-2: Health Wagon's Funding Flywheel.

free specialist consults to rural residents without needing to fund full-time specialty staff. Lab costs are offset through academic partnerships and pharmaceutical patient assistance programs help cover medications for uninsured patients. The model reveals a core truth: when systems intentionally remove profit-seeking structures, a tremendous amount of care can be delivered for very little money (see Diagram 8.2).

> **When systems intentionally remove profit-seeking structures, a tremendous amount of care can be delivered for very little money.**

While the Health Wagon also scores high on the **Physical Pillar** (mobile units to remote areas), the **Cultural Pillar** (staff from the community), the **Digital Pillar** (telemedicine integration), and the **Trust/Knowledge Pillar** (decades of reputation), it is the financial reengineering that enables all the rest. What the Health Wagon demonstrates most powerfully is that cost does not have to be a gatekeeper to quality. Instead, cost can be recast as a design constraint;

one that, when embraced, fosters innovation rather than scarcity. In a healthcare landscape often dominated by margin-driven decisions, the Health Wagon stands as a quietly radical counterexample.

How Market Forces Shape Digital Health Adoption

Even the most revolutionary healthcare technologies can fail if they are too expensive, poorly distributed, or lack regulatory approval. The story of digital health isn't just about invention; it's about affordability, accessibility, and policy alignment. Unitaid, a global health initiative in existence for the last 15+ years, has recognized that a multitude of barriers were preventing widespread access to needed tests, medications, and other products for everything from HIV to TB and malaria.[8] Without intervention, these life-saving products would remain financially out of reach for millions. By pooling funds, negotiating bulk pricing, and working with local governments, Unitaid successfully drove down costs and expanded access, making these healthcare products a practical reality for people across low-resource nations around the world.

A similar model has been used by the Medicines Patent Pool (MPP), which secures voluntary patent licenses from major pharmaceutical companies, allowing for the mass production of affordable generic drugs. MPP was created in 2010 by Unitaid. While their focus is on treatments for HIV and Hep C, they expanded their charter during the COVID-19 pandemic, working to gain access to antiviral treatments, ensuring that developing nations weren't left behind in global distribution efforts.[9]

Meanwhile, the Clinton Health Access Initiative (CHAI), serving nearly 35 countries, helps governments maximize their impact by negotiating agreements with the private sector to make their health products more affordable. They help countries set up treatment protocols and laboratory systems to do testing, and they supply systems and training for healthcare workers. They also focus on HIV but have broad reach into malaria and even contraceptives. By negotiating price

---

[8] Unitaid. *Unitaid Strategy 2023 2027*. Unitaid website. Published 2022. Accessed May 21, 2025. https://unitaid.org/uploads/Unitaid_Strategy_2023-2027.pdf.

[9] Medicines Patent Pool. *Impact*. Medicines Patent Pool website. Accessed May 21, 2025. https://medicinespatentpool.org/progress-achievements/impact.

reductions and working with supply chains to improve distribution, CHAI helps ensure that essential medications reach the people who need them most, quickly and affordably.[10]

Ensuring inclusive access to digital health services often requires going beyond conventional government budgets and donor funds. Unitaid, for instance, derives a significant portion of its revenue from an airline tax in France, creating a predictable funding pool.[11] GAVI, another player in this space, focuses on a vaccine procurement strategy that employs bulk purchasing and diversified supply chains to make vaccines more affordable and consistently available in low-income regions, an approach that has significantly improved global immunization coverage.[12]

Public-private partnerships (PPPs) are another critical vehicle for funding digital health programs, especially in low-resource settings. When public-sector oversight combines with private-sector efficiency, the result can be faster deployment of telemedicine, AI-driven diagnostics, and other digital tools. These collaborations reduce financial risks, align stakeholder incentives, and help drive innovations like mobile health services, electronic health records, and teleconsultation platforms to scale.

What these organizations reveal is a fundamental truth: innovation does not guarantee accessibility. Without deliberate market interventions—through price negotiations, licensing agreements, or financial incentives—even the most advanced technologies can fail to reach the people they are meant to serve.

Fact Check: Medicines Patent Pool By the Numbers[13]

1. 43.56 billion doses of treatment supplied to low- and middle-income countries, January 2012 through December 2023.
2. 170,000 deaths averted by the end of 2030.

---

[10] Clinton Health Access Initiative. *Market Shaping*. CHAI website. Accessed May 21, 2025. https://www.clintonhealthaccess.org/our-programs/market-shaping/.

[11] Unitaid. *Unitaid Strategy 2023 2027*. Unitaid website. Published 2022. Accessed May 21, 2025. https://unitaid.org/uploads/Unitaid_Strategy_2023-2027.pdf.

[12] Gavi, the Vaccine Alliance. Gavi website. Accessed May 21, 2025. https://www.gavi.org/.

[13] Medicines Patent Pool. *Impact*. Medicines Patent Pool website. Accessed May 21, 2025. https://medicinespatentpool.org/progress-achievements/impact.

3. 148 countries benefited from access to MPP-licensed products.

4. Estimated US $3.9 billion saved by the global community through 2030.

Value Creation and Cost Avoidance

Digital access programs create value in two ways at once: they extend clinical reach and remove avoidable cost. The Kenyan consortium that linked Kenyatta University Teaching, Referral, and Research Hospital with three county sites in 2024 shows the pattern clearly. Riding Safaricom's new 5-G backbone, a shared telesurgery console now enables Nairobi specialists to guide complex abdominal and urologic procedures while patients stay close to home.[14] Earlier Kenyan telemedicine pilots had already demonstrated high-diagnostic concordance and meaningful patient cost avoidance,[15] and global reviews confirm that digital-first programs in low- and middle-income settings routinely erased greater than 70 percent of travel-related expense.[16]

While travel costs swallow Kenyan shillings, an even quieter drain eats at every American premium dollar. Only 80 cents of each private-insurance dollar buys actual care; the other 20 cents disappears into administration, commissions, taxes, and margin (see Diagram 8.3).[17,18] That hidden surcharge is now the

> **Only 80 cents of each private-insurance dollar buys actual care.**

---

[14] Safaricom PLC. *Safaricom annual report 2024: expanding a 5-G backbone across 43 counties.* Nairobi, Kenya; Published July 2024. Accessed May 22, 2025. https://www.safaricom.co.ke/annualreport_2024/wp-content/uploads/2024/07/safaricom-annual-report.pdf.

[15] Qin R, Dzombak R, Amin R, Mehta K. Reliability of a tele-medicine system designed for rural Kenya. *J Prim Care Community Health.* 2013;4:177–181.

[16] Picozzi P, Nocco U, Puleo G, Labate C, Cimolin V. Telemedicine and robotic surgery: a narrative review to analyze advantages, limitations and future developments. *Electronics.* 2024;13(1):124. https://doi.org/10.3390/electronics13010124.

[17] America's Health Insurance Plans (AHIP). *Your Health Care Dollar: Vast Majority of Premium Pays for Prescription Drugs and Medical Care.* Press release. Published September 6, 2022. Accessed May 22, 2025. https://www.ahip.org/news/press-releases/your-health-care-dollar-vast-majority-of-premium-pays-for-prescription-drugs-and-medical-care.

[18] Centers for Medicare & Medicaid Services. *Medical Loss Ratio (MLR) Fact Sheet.* Updated August 2024. Accessed May 22, 2025. https://www.cms.gov/marketplace/private-health-insurance/medical-loss-ratio.

single biggest **Financial Barrier** to digital transformation on U.S. soil.

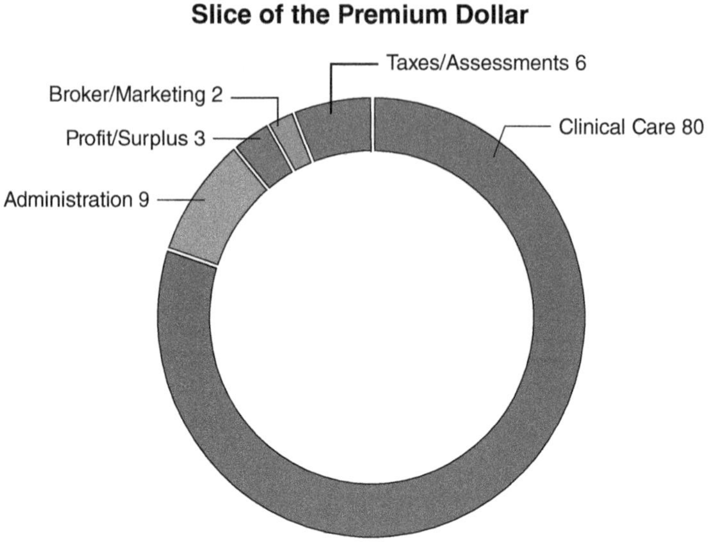

Slice of the Premium Dollar

Diagram 8-3: Where the Other 20 Cents Goes.[19,20]

Those missing 20 cents act like an unseen tax on innovation: every billion siphoned into overhead is a billion not available for AI triage pilots, remote-robotic beta sites, or sensor-driven home recovery programs (see Diagram 8.4).

Each row of the table in Diagram 8.4 translates a stubborn cost line in the employer claim file into a pointed question and an evidence-based negotiating point. Administration comes first. Best-in-class, self-funded plans keep administrative-services-only (ASO) fees at about 5 percent of the premium. ASO fees are made up of third-party administrator charges to process claims, adjudicate eligibility, and costs to issue estimation of benefits (EOBs).

---

[19] America's Health Insurance Plans (AHIP). *Your Health Care Dollar: Vast Majority of Premium Pays for Prescription Drugs and Medical Care.* Press release. Published September 6, 2022. Accessed May 22, 2025. https://www.ahip.org/news/press-releases/your-health-care-dollar-vast-majority-of-premium-pays-for-prescription-drugs-and-medical-care.

[20] Centers for Medicare & Medicaid Services. *Medical Loss Ratio (MLR) Fact Sheet.* Updated August 2024. Accessed May 22, 2025. https://www.cms.gov/marketplace/private-health-insurance/medical-loss-ratio.

LINE ITEM	ASK THIS	SAMPLE TACTIC
Administration	What percent of my premium is non-clinical?	Demand ASO fee transparency; benchmark vs 5% best-in-class
Network Rates	How do unit prices compare to Medicare?	Negotiate reference-based pricing for commoditized imaging
Pharmacy	Are PBM spreads disclosed?	Opt for pass-through PBM with lowest net cost guarantee
Innovation Fund	How much of savings recycle into new care models?	Ring-fence 25% of achieved savings for virtual care expansion

Diagram 8-4: Toolbox: Reclaiming the Missing 20 Cents.

Requiring a transparent ASO schedule and benchmarking it against that threshold exposes avoidable overhead.[21]

Network rates follow. Asking how unit prices compare with Medicare equips purchasers to pursue reference-based pricing (RBP) for high-volume, commodity care. Vendor-based RBP programs typically reduce employers' overall medical spend by 20–30 percent. However, there are pros and cons to reference-based pricing, and risks include a loss of protection since there are no provider contracts.[22]

Pharmacy spend is next. Here, the pivot is the pharmacy benefit manager (PBM), the intermediary that negotiates drug prices, sets formularies, and administers rebates. Whether a PBM uses "spread pricing" (buying low, billing high) or passes manufacturer rebates straight through to the plan sponsor determines how much waste lingers in every prescription dollar. A large percentage of employers don't currently receive a full rebate pass-through. With all the complexities it can seem overwhelming, but there are many ways to accelerate payback.

---

[21] NASCO. *Does your health plan meet the needs of ASO employers?* Published March 21, 2024.

[22] National Alliance of Healthcare Purchaser Coalitions. *A Fresh Look at Reference-Based Pricing to Drive Affordability.* Published 2024. Accessed July 14, 2025. https://www.nationalalliancehealth.org/resources/a-fresh-look-at-referenced-based-pricing-to-drive-affordability/.

Next Shift Quick Wins: Three Levers That Accelerate Payback

LEVER	HOW IT WORKS	WHY IT MATTERS
Mobile-money reimbursement	Insurers push claim payments to M-TIBA wallets within 72 hours	Program audits show a 40% drop in cash leakage and a rapid uptick in enrollment[23]
Shared cloud analytics hub	County hospitals pool storage, analytics, and security in one subscription	Cuts per-facility capital and opens peer-to-peer quality dashboards without new staff
Community Wi-Fi telehealth kiosks	Telco sponsors bandwidth at market stalls and bus depots, where nurses book follow-ups	CMS Hospital at Home pilots report a 14% lift in visit completion and fewer no-show write-offs[24]

The three quick wins each shave time or waste out of the cash cycle. First, routing insurer reimbursements straight into M-TIBA mobile wallets cuts the claims-payment lag to 72 hours. The sudden plug in cash leakage sparks faster patient enrollment and pulls the break-even point forward by about three months. Second, a shared cloud-analytics hub lets the county hospitals rent a single stack of storage, security, and analytics licenses, trimming capital expenditure and nudging the curve upward another two months. Finally, free community Wi-Fi kiosks at market stalls and bus depots carry virtual-care bandwidth to where patients actually live. Follow-up completion rates jump 14 percent, turning no-shows into revenue and erasing the last month of deficit. Layered together, these quick wins turn a respectable payback into a self-financing flywheel.[25]

[23] America's Health Insurance Plans (AHIP). *Your Health Care Dollar: Vast Majority of Premium Pays for Prescription Drugs and Medical Care.* Press release. September 6, 2022. Accessed May 22, 2025. https://www.ahip.org/news/press-releases/your-health-care-dollar-vast-majority-of-premium-pays-for-prescription-drugs-and-medical-care.
[24] Centers for Medicare & Medicaid Services. *Medical Loss Ratio (MLR) Fact Sheet.* Updated August 2024. Accessed May 22, 2025.
[25] Matt S. No copays, no problem: how one mobile clinic survived without a billing department or a therapist for their grant writer. *Unpublished but emotionally validated.* 2025.

Partnerships, Pilots, and the Price of Persistence

When capital is tight, innovation doesn't disappear, it just gets quieter. It becomes a series of backroom conversations, joint venture pivots, and slow-burning pilots that never make headlines. These are the small bets that depend less on vision decks and more on trust between partners. But even the smartest models can collapse if their funding dries up before they mature. In this next vignette, Danielle Church, JD reflects on the fragile math behind digital transformation, and explains how real change often hinges less on brilliance than on budgeting discipline.

Stories from the Field: Financial Barrier Spotlight[26]

Danielle Church JD, MBA has spent her career in the thick of health system finance; spanning payer strategy at Aetna, retail innovation at CVS Health, and joint-venture delivery reform. If anyone knows what happens when financial incentives don't line up, it's her. And the word she uses most often to describe the challenge? Misalignment.

"We keep trying to innovate in systems built to say 'no'," Danielle reflected. From health hubs in Houston to risk-sharing physical therapy ventures in Seattle, to observing CVS acquisitions like Oak Street, she has seen ideas spark, stall, and sometimes disappear altogether. Not because the clinical idea was wrong, but because the financial math didn't hold long enough for real change to take root.

Her insight echoes a core financial truth: in healthcare, every project competes with every other project. Even for hospital systems that want to modernize care, short-term cost pressures often derail long-term bets. "You can't just bolt innovation onto the side of a financially strapped organization," she said. "Everyone's working with a shrinking pot, and even within a single system, what matters most to one leader may be a rounding error to another."

At ATI, the national physical therapy group she advises, Danielle has seen a different model take shape: joint ventures with hospitals that offload their outpatient rehab units. The upside? ATI runs the clinics as a core service, not a line item. The hospital reduces overhead, ATI gets access to better payer rates, and patients see improved MSK (musculoskeletal) continuity. But even that model faces headwinds.

---

[26] Matt S. Interview with Danielle Church, JD, Healthcare Consultant and Former Strategy Leader, CVS Health. June 2, 2025.

"Physical therapy is not high-margin, and in a tight year, it's the first thing to get deprioritized."

Danielle believes that what works best is often what's narrowly defined. "Organizations like Oak Street didn't try to do everything. They stayed tightly focused on primary care, in specific geographies, with a clear reimbursement model." That constraint, she argues, became their competitive advantage. "Healthcare strategy only works when it's simple enough to fund and sustain."

What kills innovation, she notes, is rarely bad intent. It's a mix of leadership turnover, shifting priorities, and the reality that most organizations must show short-term gains every quarter. "We all want long-term transformation," she says. "But we fund like we're only trying to survive the next 12 weeks."

Her closing advice? Financial innovation must be treated like clinical innovation: measured, managed, and protected. Otherwise, healthcare ends up full of "maybes," programs that could have worked, if only the dollars had held out.

Stories from the Field: Financial Barrier Spotlight[27]

Shalu Gugnani, MD spent eight years building a full continuum addiction program at Northwest Community Healthcare in suburban Chicago. Early in that work she met Marisa, a 43-year-old project manager who finally agreed to 30-day residential treatment after two opioid-related admissions. On admission day, the utilization-review fax revealed that her commercial plan would authorize only seven days, pending "acute psychiatric reassessment." That single line detonated months of preparation. Marisa calculated the uncovered balance, about $19,000, over half of the $26,000 national median for a month of residential care.[28] Unable to pay, she left against advice on day eight. Within 10 days, she relapsed, returned through the emergency department, and entered the same unit for detoxification. By discharge, her insurer had paid for two separate acute hospital stays, yet never covered the uninterrupted course her clinicians had recommended.

Dr. Gugnani calls this pattern "insurance friction." Reviewers often mislabel stable addiction as an acute crisis. If the patient is not suicidal,

---

[27] Matt S. Interview with Shalu Gugnani, MD. March 26, 2025.
[28] Residential addiction treatment for adolescents is scarce and expensive. *National Institutes of Health News Release*. Published January 8, 2024. Accessed May 16, 2025. https://www.nih.gov/news-events/news-releases/residential-addiction-treatment-adolescents-scarce-expensive.

coverage vanishes. This is a loophole that violates the spirit, though not always the text, of the 2008 Mental Health Parity and Addiction Equity Act.[29] She has found that in peer-to-peer appeals, she succeeds roughly half the time. In the other half, patients shoulder either a sudden bill or a forced discharge, both of which amplify relapse risk and long-term cost. For those who self-pay, her hospital required roughly 51 percent of total cost upfront (often $13,000 or more), an impossible sum for many families.

COVID-19 loosened certain barriers. Telehealth follow-ups let employed patients alternate in-person and virtual visits. Marisa herself later relied on hybrid appointments to keep a new job. Yet technology could not erase the front-end cash shock, nor the stress of daily utilization reviews that ask patients to prove "sickness" while learning recovery.[30] "Parity laws mean nothing when the reviewer still treats addiction as a choice," Dr. Gugnani tells her trainees, recounting how she once blurted "you are a bad person" at a reviewer who insisted detox was the only medically necessary service. This statement really touched me, and I like Dr. Gugnani even more because of it!

Downstream costs escalate silently. Dr. Gugnani shared that generally every dollar spent on substance-use treatment yields seven dollars in societal benefit, mainly by averting emergency care and criminal justice expenses. Premature discharge squanders that return. Marisa's story illustrates the paradox: rigid utilization controls intended to save money created a revolving-door pattern that ultimately cost her insurer more while jeopardizing her life.

For innovators designing value-based payment models, the lesson is clear; any mechanism that withholds coverage until crisis emerges contradicts both clinical evidence and financial prudence. Capital must arrive early, stay predictable, and align with recovery milestones, or the business case for addiction care collapses along with the patients it is meant to serve. When we debate bundled payments and risk corridors in the sections that follow, we should recall Marisa's unfinished ledger—seven authorized days, 23 denied, two readmissions, one life still in limbo.

---

[29] Substance Abuse and Mental Health Services Administration. *Know Your Rights: Parity for Mental Health and Substance Use Disorder Benefits.* Publication ID PEP21-05-00-003. Published April 2022. Accessed May 16, 2025. https://library.samhsa.gov/product/know-your-rights-parity-mental-health-and-substance-use-disorder-benefits/pep21-05-00-003.
[30] Ibid.

Payment Models Reimagined

Dr. Gugnani's experience caring for patients with substance-use disorders lays bare a central truth: when coverage is sliced into scattered authorizations and short approvals, clinicians spend more time fighting denial letters than treating disease. Patients, meanwhile, step through a revolving door of relapse and readmission, unsure who will pay and for how long. This friction at every hand-off reminds us that access is not only a product of geography or technology, it is also a function of the financial plumbing beneath each clinical encounter. With that reality in view, we now turn to payment models, a fresh look at how bundled episodes and prospective rates can replace reactive billing with anticipatory care planning, shifting the question from "who will approve this visit" to "how do we invest early so patients never reach crisis."

In the United States, the Center for Medicare and Medicaid Innovation now uses the Transforming Episode Accountability Model (TEAM) to pay one prospective rate that covers the index hospital stay, any qualified hospital at home days, remote physiologic monitoring, infusion, and respite nursing. The single bundle runs through the 30-day recovery window and shifts clinical planning to the moment of admission rather than the day of discharge.[31,32]

Early evidence from hospital-at-home programs folded into similar prospective bundles shows a 26 percent relative reduction in 30-day readmissions when compared with traditional inpatient management.[33] Virtual follow-up further extends the benefit. In an academic stroke clinic, attendance rose from 19 to 41 percent once telemedicine visits were offered, a gain that doubled completion of recommended reviews.[34] TEAM also layers an equity accelerator. Hospitals may file

---

[31] Centers for Medicare & Medicaid Services. *Transforming Episode Accountability Model Fact Sheet*. Published April 10, 2024. Accessed May 23, 2025. https://www.cms.gov/files/document/team-model-fs.pdf.

[32] Bailey V. Breaking down the new CMS proposed bundled payment model: TEAM. *RevCycleIntelligence* website. Published May 1, 2024. Accessed May 23, 2025. https://www.techtarget.com/revcyclemanagement/news/366600169/Breaking-down-the-new-CMS-proposed-bundled-payment-model-TEAM.

[33] Arsenault Lapierre G, Henein M, Gaid D, Le Berre M, Gore G, Vedel I. Hospital at home interventions versus in hospital stay for patients with chronic disease who present to the emergency department: a systematic review and meta analysis. *JAMA Netw Open*. 2021;4(6):e2111568. Accessed May 23, 2025. https://pmc.ncbi.nlm.nih.gov/articles/PMC8188269/.

[34] Alabyad D, Lemuel-Clarke M, Antwan M, Henriquez L, Belagaje S, Rangaraju S, Mosley A, Cabral J, Walczak T, Ido M, Hashima P, Bayakly R, Collins K, Sutherly-Bhadsavle L, Brasher C, Danaie E, Victor P, Westover D, Webb M, Skukalek S, Barrett AM, Esper GJ, Nahab F. Telemedicine impact on post-stroke

voluntary health equity plans and receive social risk adjustment. There has been additional discussion on incentives that would subsidize connectivity for rural beneficiaries so that lack of broadband does not become a new barrier to recovery in the home setting. See Diagram 8.5.

Diagram 8.6 reads like a four-lane roadmap that moves from fee-for-service at one end to full capitation at the other. Each stop on that road lights up more of the five access pillars. Fee-for-service leaves most of the bridge dark because hospitals are paid only when they deliver a test or a visit. Patients still travel, pay coinsurance, and hope their records follow them. TEAM pushes the light farther by paying for hospital-at-home days, remote monitoring, and respite nursing in one bundle. Money follows the patient through the first month at home and lets teams invest in tablets and visiting nurses that keep the **Physical, Digital**, and **Financial Pillars** open. Finally, a high per member per month budget turns on every bulb. Plans that live on a steady monthly rate adopted tele-visits fastest during the pandemic and even mailed broadband kits and food cards to members, steps that also build the **Cultural** and **Trust/Knowledge Pillars**. The diagram shifts from a narrow beam to full floodlights as you move right, showing how broader payment responsibility unlocks wider access.

Building a Sustainable Infrastructure: Local Challenges Require Local Solutions

Global improvements often require very specific local enablers. For instance, electricity and connectivity remain serious hurdles in rural areas. In some communities, acceptance of technology might hinge on trusted individuals, like community health workers or nurses, acting as intermediaries **(Cultural Barrier)**. Many of the most successful case studies involve government or NGO (nongovernmental organizations) support to subsidize these programs, reflecting the fact that technology alone isn't a cure-all. Local buy-in, sustained funding, and seamless system integration are equally important. Implementing a digital health program successfully over the long-term differs drastically from merely launching one. Ongoing training, tech support, reliable Internet connections, and a supportive policy environment determine whether solutions will thrive or merely disappear after the pilot phase.

outpatient follow-up in an academic healthcare network during the COVID-19 pandemic. *J Stroke Cerebrovasc Dis*. 2023 Aug;32(8):107213. doi:10.1016/j. jstrokecerebrovasdis.2023.107213. Epub 2023 Jun 21. PMID: 37384981; PMCID: PMC10284452.

PAYMENT MODEL	CASHFLOW PREDICTABILITY (PROVIDER)	OUT-OF-POCKET (OOP) PREDICTABILITY (PATIENT)	LEVER FOR ACCESS PILLARS	NOTABLE CMS EXAMPLE	NET EFFECT ON FIRST-YEAR GEOGRAPHIC REACH
Fee-for-service	Low; volume fluctuates	Low; bills post-care	Weak on **Physical Pillar**; no guardrails	Traditional Medicare Part B	Baseline
Capitation[35]	High; monthly per member per month	High; flat copay/none	Strong **Financial Pillar**; mixed **Cultural Pillar**	Primary care first capitation track	+12%
Shared savings (ACO), DRG (Diagnosis Related Group)	Moderate; upside corridors	Moderate; no OOP caps	Adds **Trust/Knowledge Pillar** via coordination	Medicare Shared Savings Program	+9%
Outcome-based/bundles[36]	High once ramped	Moderate; capped per episode	Balanced across pillars; equity bonuses	BPCI-advanced surgical bundles; TEAM	+15%

Diagram 8-5: Impact of Four Payment Models on Access Indicators.

[35] Association between capitated payments and preventive care delivery. *JAMA Netw Open.* 2023;6(12):e234556. Accessed May 13, 2025. https://pubmed.ncbi.nlm.nih.gov/3779068/.

[36] Centers for Medicare & Medicaid Services. *Bundled Payments for Care Improvement Advanced Model.* Accessed May 13, 2025. https://innovation.cms.gov/innovation-models/bpci-advanced.

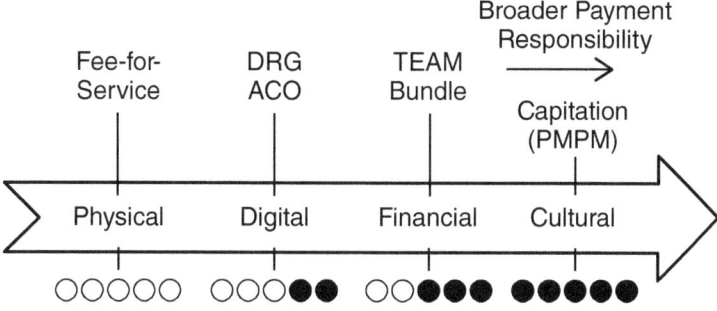

Diagram 8-6: Payment Model Roadmap to Access.

Stories from the Field: Financial Pillar Spotlight[37]

Singapore's HealthHub began not as a standalone app, but as a public capital strategy: a blueprint to make digital health services accessible to every resident, sustainably. The platform was one of the earliest major projects under Singapore's Smart Nation initiative and was designed with three **Financial Pillar** access goals from the start:

- Remove cost **(Financial Barriers)** for patients and caregivers.
- Reduce infrastructure duplication across agencies and vendors.
- Shift long-term platform costs off public balance sheets without compromising quality or citizen coverage.

Each goal was met by pairing cloud infrastructure with smart public spending. Instead of building bespoke data centers or contracting each agency independently, Singapore moved to a whole-of-government cloud model and used HealthHub to demonstrate how pooled architecture and cost recovery could make digital health both universal and economically viable. Singapore pooled agency infrastructure spend into a whole-of-government cloud, reducing duplication and enabling self-service use cases without increasing access cost.

---

[37] AWS Institute. *How the Cloud Can Help Transform Health Systems.* Amazon Web Services; 2021:48–52. Accessed June 4, 2025. `https://d1.awsstatic.com/institute/Cloud%20for%20Healthcare Overview.pdf`.

At launch, Singapore's government absorbed the full cost of HealthHub's design, hosting, and integration. The system was hosted in the Government Commercial Cloud (GCC), which provided scalable, shared infrastructure for all participating agencies, including the Ministry of Health, the Health Promotion Board, polyclinics, and public hospitals.

This decision dramatically lowered the startup cost for each partner. APIs for lab results, prescriptions, immunizations, and appointment bookings were built once and reused across the system, with no vendor-side replication costs. A shared authentication layer (Singpass) allowed citizens to log in without new accounts or passwords, eliminating access friction.

Access insights include **(Financial Pillar)**:

- Patients paid nothing to access services.
- Partners avoided upfront infrastructure and licensing fees.
- Common APIs reduced the total cost of ownership across government agencies.
- Public investment focused on interoperability, not on branded software.

Once the core platform stabilized, Singapore began gradually shifting HealthHub's operating costs to its ecosystem partners. Using usage-based pricing, ecosystem partners—commercial labs, private providers, and insurers—paid transaction-based fees for accessing shared APIs through the platform. These charges were designed not as profit mechanisms, but as cost-recovery tools that scaled with the value delivered.

The system's cloud-native backend made this possible. APIs were metered and access-controlled by design, allowing Singapore to introduce fine-grained subscription and pay-per-use models. These weren't pay-per-use models for patients, but for institutional partners who processed claims, issued prescriptions, or retrieved diagnostic results through the platform. The usage-based funding model allowed HealthHub to remain free at the point of use while shifting operational costs to those who financially benefit from the transactions.

By 2020, approximately 70 percent of HealthHub's running costs were covered through these commercial usage charges, with the remaining 30 percent funded by the Ministry of Health and the Health Promotion Board.

Access insights include **(Financial Pillar)**:

- Cloud architecture-enabled metered pricing without disrupting service.
- Cost recovery focused on B2B API calls, not user-side charges.
- Institutional partners paid in proportion to their system use.
- Free citizen access throughout.

HealthHub's financial model was only possible because of its cloud-first technical design. Rather than building individual platforms for individual services, Singapore centralized its digital health backbone in the government cloud and invested in system-wide identity, interoperability, and API tooling. This created economies of scale that made cost-sharing feasible, and universal access defensible.

Because every agency and vendor shared the same cloud foundation, recurring costs could be predicted and distributed transparently. Because data was standardized across the system, new services could be added with low incremental cost. And because authentication and messaging were unified, residents could access everything from COVID-19 test results to prescription refill payments, using a single login and mobile interface, regardless of income or provider.

Access insights include **(Financial Pillar)**:

- Cloud scale kept marginal service costs low.
- Open API ecosystem encouraged private-sector adoption.
- Platform economics supported cashless, centralized service delivery.
- Free usage for residents was protected even as service complexity grew.

HealthHub's financing model is not about a single innovation. It is a systemic blueprint: anchor a public platform in shared cloud infrastructure, meter usage by institutional partners, and use that revenue to fund ongoing delivery. Singapore showed that when infrastructure is pooled and access is centrally authenticated, cost-sharing becomes not only viable, but beneficial to all parties. While this is an example based on AWS, the major cloud hyperscalers, including OCI and Azure, can also meet country-wide cloud needs of this type. When I was at OCI, these are the types of transformations we drove everyday with Ministries of Health and large health systems around the world.

This model can be adapted by other nations looking to protect patient access while building financially durable digital services.

While many of the case studies in this chapter emerged from top-down budget shifts or national stimulus, similar financial access levers can be applied at the hospital, network, or payer level. Diagram 8.7 maps real-world financial tactics to their dominant and secondary access pillars, which can help leaders turn broad policy goals into quantifiable outcomes.

ACTION	FINANCIAL PILLAR VALUE LEVER	SECONDARY PILLAR
Balance sheet first rounds	Unlock capital by shifting early spend to inexpensive pilots	Digital Pillar
CapEx to OpEx swap	Convert hardware buys into subscription services	Digital Pillar
Time value of data	Monetize real-time clinical data flows for pay-for-performance	Trust/Knowledge Pillar
Liquidity via shared infrastructure	Pool regional demand to finance cloud and fiber backbone	Cultural Pillar
Marginal cost lens on scaling	Apply continuous cost curve analysis before expansion	Physical Pillar
Risk premium cut with zero-trust	Lower insurance spend through security hardening	Trust/Knowledge Pillar
ROI tracking dashboard	Embed cash and outcome metrics in monthly board pack	Digital Pillar
Outcome credits pooling	Bank financial gains from reduced readmissions to fund innovation	Digital Pillar
Cost of downtime audit	Quantify revenue losses per minute of system outage	Digital Pillar
Cross-subsidy innovation fund	Earmark portion of profitable service lines for access projects	Cultural Pillar

Print this page and pin it to your war room wall.

Diagram 8-7: Ten High Impact Moves That Strengthen the Financial Pillar While Activating At Least One Additional Access Pillar.

Risk, Governance, and Return

Access is often framed as a moral issue or a logistical one, but it is also a question of capital. Who holds it, who allocates it, and who is willing to risk it. For all the language around innovation, the healthcare system still hesitates to invest in the messy, unscalable business of human connection. We chase high-yield technologies and scalable platforms, but flinch when the return on investment cannot be easily modeled. Yet some of the most transformational results come not from capital-intensive systems redesign, but from relational risks taken at the edge of the system. Sometimes, the most valuable infrastructure is a folding chair on a sidewalk. The story of Dr. David Lehmann is a case in point.

Stories from the Field: Trust/Knowledge Pillar Spotlight[38]

David Lehmann, MD met Jose, a 40-year-old man with endocarditis, after Jose had entered the emergency department 27 times in 60 days. Every visit was billable, yet none solved the underlying problem. As you've seen in previous chapters, Lehmann quit fulltime hospital practice, partnered with community group In My Father's Kitchen, and began offering primary care on the sidewalk from a mobile unit. Over the next six months, his street medicine team followed 50 chronically unsheltered patients. Their emergency revisits consequently fell by one half and inpatient days dropped by one quarter. Lehmann insists the savings flowed not from new hardware but from relocating first contact to the curb. Trust, not technology, unlocked return on investment. "Medicaid paid every bill; what the system lacked was a doctor who would answer when the curb called," he explains. The lesson for boards is clear—capital allocation that ignores relational trust leaves money on the table as surely as it leaves patients on the street.

While government spending and NGO partnerships have shaped huge swaths of the industry's growth, where are the private investors doubling down? This is an interesting way to consider the market growth and its implications for access. As you have learned, without sustainable funding from all sources, new health innovations can't survive.

Digital health funding in the United States slowed only a notch in 2024 with US $10.1 billion over 497 deals. This still cleared the

---

[38] Matt S. Interview with David Lehmann, MD, and Mia Ruiz-Salvador. March 20, 2025.

pre-pandemic bar, yet investors rewired the mix away from consumer apps toward fewer "bling apps" and more backbone infrastructure. Rock Health's year-end scan shows that nonclinical workflow and data-plumbing startups investment were up, second only to therapeutic point solutions, while AI-enabled tools, often embedded in those same pipes, captured 37 percent of all capital.[39]

The purse strings of provider organizations finally opened in 2024. A Bain & Company survey of 150 U.S. systems and payers reports that three quarters increased IT budgets in 2024 and ranked infrastructure, cybersecurity, and interoperability as their top spend targets.[40] Boards that wrestled with the Change Healthcare outage now treat data pipes and payment rails as clinical throughput rather than backoffice cost centers. A new necessity, and a new, "never event."

Where the Numbers Meet the Nerve

Money was never meant to headline this book, yet it keeps crashing the party. Chapter 1 explained that *presence* launches access; by Chapter 4, you saw how *culture* decides whether that front door ever opens. Here, deep in balance-sheet acrobatics and razor-thin margins, is a quieter truth—finance is the hinge that can swing every other pillar wide or bolt them shut.

Budgets behave like blood vessels: constricting in one region, dilating in another, always affecting the whole body. Singapore's cloud consortia, new funding models, Danielle Church's creative carve-outs—each story shows the **Financial Pillar** as both constraint and catalyst. It turns Wi-Fi into a surgical lifeline, a community bond into broadband, and a line item into a lived reality.

Yet finance never works alone. A zero-trust security upgrade (Chapter 7) collapses without clinician trust. A gleaming remote-surgery robot (Chapter 5) gathers dust if the reimbursement code stalls. The pillars are interdependent load-bearers: when one shifts, the others must compensate or the whole structure groans. Every invoice is also a cultural artifact, every grant proposal a digital roadmap, every margin target a silent vote on patient wellbeing.

---

[39] Rock Health. 2024 year-end market overview: Davids and Goliaths. Published January 13, 2025. Accessed May 23, 2025. https://rockhealth.com/insights/2024-year-end-market-overview-davids-and-goliaths/.
[40] Berger E, Dowling C, Feinberg A, Hammond R. Healthcare IT spending: innovation, integration, and AI. *Bain & Company*. Published September 17, 2024. Accessed May 23, 2025. https://www.bain.com/insights/healthcare-it-spending-innovation-integration-ai/.

So where do we go from here? Chapter 9 dives into quality in the digital age: real-time dashboards, algorithmic bias checks, and the messy human work of turning data into safer, kinder care. We'll test whether the financial scaffolding we've explored truly supports excellence or merely balances the books. Chapter 10 looks farther ahead, asking how tomorrow's technologies, payment models, and societal expectations might redraw every pillar built so far. Because if eight chapters have taught us anything, it's this: access is a composite art, and finance is the brush that can either outline, or erase, the future we're all trying to paint.

Turn the page; the ledger has handed the pen to quality, and the future is waiting to co-sign.

🔧 Startup Builder's Box: Economics Survival Guide

FUTURE PROOFING	WHAT TO DO NOW!
1. **Pick the cash leak you plug first.**	Show the CFO an explicit cost line, such as avoidable readmissions or manual prior-authorization runs, that your product will shrink within 12 months. Translate that impact into margin per case rather than abstract return on investment to win budget priority.
2. **Design for the payment model that writes the check.**	Package your solution so it fits inside prospective bundles like TEAM and can feed the equity accelerator with social-risk data fields that hospitals must now report. This alignment moves you from "nice to have" to "required line item" when the bundle launches.
3. **Prove lift on at least one access pillar beyond the Financial Pillar.**	Capture a clear metric (fewer miles traveled, lower out-of-pocket exposure, or higher virtual-visit completion) and publish it in a one-page infographic that your champions can circulate internally.
4. **Build your "budget slide" with the numbers investors track.**	Venture funding in digital health hit US $10.1 billion across 497 deals in 2024, with 63 percent of rounds labeled seed through Series B. Early-stage capital is still flowing, but large buyers dominate scale-up exits, so show a credible partnership path with health-system Goliaths.[41]

(continues)

---

[41] Rock Health. 2024 year-end market overview: Davids and Goliaths. Published January 13, 2025. Accessed June 23, 2025. https://rockhealth.com/insights/2024-year-end-market-overview-davids-and-goliaths/.

🔧 **Startup Builder's Box: Economics Survival Guide (continued)**

FUTURE PROOFING	WHAT TO DO NOW!
5. Match the enterprise mood on tech spend.	Seventy-five percent of U.S. providers and payers increased IT budgets last year and cite rapid return and seamless integration as gating criteria. Your demo must prove "install in days, deliver savings in weeks" and should include a sandbox build that drops into the client's existing cloud identity layer.
6. Ship with compliance guardrails baked in.	Bundle the legal memo that maps HIPAA clauses, clinical decision support rules, and emerging algorithm-transparency standards. Startups that remove regulatory friction score faster security reviews and shorten contracting cycles.
7. Hard-wire add-on metrics.	Expose dashboards that track usage by rural ZIP code, preferred language, or disability status so customers can claim TEAM social-risk add-ons and community-benefit credits without extra analytics spend.
8. Plan for 24-month runway, minimum.	Payment-model pilots often delay checks for six months while reconciliation algorithms mature. Secure bridge capital or phased-implementation fees so you are not starved at the point customers need proof.
9. Court the clinical-finance duo.	Win both the CMO and the revenue-cycle lead. Startups that anchor in only one office stall when procurement gates open. Schedule joint read-outs every quarter with metrics tied to each leader's scorecard.
10. Document every avoided cost in real time.	Embed claim-scrape or EHR queries that auto-populate your impact tracker. Founders who walk into renewal talks with live dashboards, not anecdotes, keep pilots alive and expand to new sites.

Leaders' End-of-Chapter Action Checklist: Chapter 8: "Economics of Borderless Care"

LEADER	HIGH-IMPACT ACTION TO STRENGTHEN AFFORDABLE CARE
❏ Board Director	Approve a Patient Financial Protection Policy that caps any household's annual responsibility at 5% of income and post yearly compliance results on the public dashboard
❏ Chief Executive Officer	Create a Cost Experience Lab within 30 days to test three affordability innovations every quarter and present impact metrics to the board
❏ Chief Financial Officer	Replace paper statements with digital wallet and text payment links within six months, aiming to cut billing overhead by 30% and redirect the savings to sliding-scale discounts
❏ Chief Information Officer	Deploy real-time eligibility checks at appointment scheduling and alert revenue-cycle leads when a patient would face more than US $500 in projected charges
❏ Chief Health Information Officer	Publish a quarterly Formulary Affordability Review, listing the 25 most commonly prescribed chronic medications and their lower-cost therapeutic equivalents
❏ Chief Strategy Officer	Secure two employer direct contracts that bundle knee or hip procedures at a transparent flat rate with travel vouchers and track patient net spend
❏ VP Revenue Cycle and Patient Financial Services	Achieve 95% pre-service financial clearance by including charity screening and flexible payment planning, then report monthly bad-debt variance
❏ VP Payer Relations	Negotiate at least one commercial contract this year that adopts the Medicare Advantage out-of-pocket maximum as an affordability guardrail
❏ Telehealth Program Manager	Open an asynchronous messaging clinic priced at about US $40 per visit and divert 10% of low-acuity follow-ups; review savings at six months
❏ Patient Financial Advocate Manager	Host quarterly Cost Clarity webinars for patients and maintain an average satisfaction rating above 4.5/5

LEADER	HIGH-IMPACT ACTION TO STRENGTHEN AFFORDABLE CARE
❏ **Community Benefit Officer**	Train 10 volunteer Health Cost Navigators to staff weekly sessions at local libraries and enroll 200 residents in subsidized coverage within six months
❏ **Director of Snacks & Morale**	Swap brand-name snacks for generics at your next team meeting; reallocate the savings to something useful (ideally tacos!)

Quality in the Digital Age

I remember back to 2013 when I started at NextGen healthcare. It was an exciting time; EHRs were still in their upward growth spurt. I came into a flurry of activity surrounding the tail end of Meaningful Use 2 (MU2) development and attestation. Meaningful Use was Washington's carrot-and-stick program to get hospitals off paper. Phase 2 turned that carrot into a 400-item checklist.

What really struck me at that time were the checkboxes. Yes, the constant checkboxes. Does it have e-prescribe capabilities ... yes, check! Does it have the ability for patients to access their summary from their visit ... yes, check! Then there were the countless quality measures. Again, so many checkboxes. If a patient has X and Y but not a full stomach on a full moon, then yes ... check! If MU2 had a love language, it was definitely, you guessed it, *checkboxes*. I was basically dating a spreadsheet.

Quality measures were being forced into medical records at "record speed," but was this actually quality? The "rules" that governed what quality looked like were well intended, but in practice the systems were becoming clunky due to the speed needed to bring features to life. Patients were none the wiser, except they kept being asked by every doctor they saw if they smoked. And why were patients being asked this question? Those providers had to check the box, since that's what their admin teams required (see Diagram 9.1).

Quality in digital health isn't just about flashy apps or fast algo-rithms. It's about delivering better outcomes, for everyone. And that means rethinking what excellence looks like in settings as diverse as a Boston academic hospital, a Nairobi clinic, and a tribal health center in Arizona. This isn't just aspirational and it's also not just a checkbox.

Diagram 9-1: Five Pillars Snapshot.

Quality in digital health is not an independent metric. It is the lived outcome when every pillar of access works in concert with the classic quality aims. The Institute of Medicine lists six aims and this chapter focuses on the five that surface most often in my interviews and case studies: clinical effectiveness, equity, safety, efficiency, and user-centeredness.[1] ("Timely" is the last element, and that goes without saying!) Each aim lands differently on the five access pillars, yet none can be achieved if any one pillar is weak (see Diagrams 9.2 and 9.3).

---

[1] Agency for Healthcare Research and Quality. Six domains of health care quality. Accessed May 26, 2025. https://www.ahrq.gov/talkingquality/measures/six-domains.html.

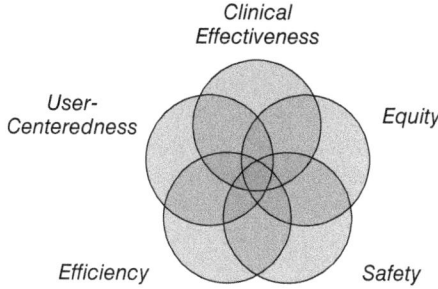

Diagram 9-2: Five Dimensions of Digital Health Quality.[2,3,4,5]

Why every pillar matters:

- If broadband drops or a portal is unreadable, clinical effectiveness stalls because updated protocols never reach frontline teams.
- When insurance design leaves patients paying first dollar, inclusive care fails even if algorithms are flawless.
- A culturally discordant tele-visit may satisfy timeliness yet still miss user-centeredness and safety because the patient leaves confused about warning signs.

The pattern is clear; quality is not a destination outside the access map. Quality is the composite score that emerges when all five pillars bare equal load. In practice, quality improvement projects should begin with a pillar check: Which pillar is weakest for this population and how will that weakness undercut the intended quality aim? Only after closing that gap do dashboards and incentives deliver the performance they promise.

---

[2] World Health Organization. *Global Strategy on Digital Health 2020–2025.* Accessed May 26, 2025. https://www.who.int/docs/default-source/documents/gs4dhdaa2a9f352b0445bafbc79ca799dce4d.pdf.

[3] National Academy of Medicine. Harnessing evidence and experience to change culture: a guiding framework for patient and family engaged care. Accessed May 26, 2025. https://nam.edu/perspectives/harnessing-evidence-and-experience-to-change-culture-a-guiding-framework-for-patient-and-family-engaged-care/.

[4] Penchansky R, Thomas J. Access to care: remembering old lessons. *Health Serv Res.* 1981;16(1):5–21. Accessed May 26, 2025. https://pmc.ncbi.nlm.nih.gov/articles/PMC1464050/.

[5] Campaign for Action. In health care, there is no quality without equity. Accessed May 26, 2025. https://campaignforaction.org/in-health-care-there-is-no-quality-without-equity/.

QUALITY AIM	PRIMARY ACCESS PILLAR(S)	HOW THE PILLAR ENABLES THE AIM
Clinical effectiveness	Digital; Trust/ Knowledge	Evidence-based guidelines and decision support reach bedside teams only when bandwidth, data interoperability, and transparent sharing of research findings exist
Equity	Financial; Cultural	Value-based payment tied to transparent pricing narrows cost barriers; language matched staff plus community champions dismantle cultural bias
Safety	Physical; Digital	Mobile outreach clinics shorten time to first contact; secure portals close the feedback loop on adverse events
Efficiency	Financial; Digital	Unified billing and data streams cut duplicate imaging and visits; tele-consults reduce idle operating time
User-centeredness	Trust/ Knowledge; Cultural	Layered transparency tools show price and quality side by side; shared decision aids in the preferred language place the patient at the design table

Diagram 9-3: Quality Aims and Their Relationship to the Pillars.

Practical Signals How-To

	PRACTICAL SIGNAL TO TRACK
Clinical effectiveness	Percentage of guideline alerts opened within two minutes of trigger
Equity	Out-of-pocket share of cost for comparable procedures across ZIP codes
Safety	Median hours from symptom onset to triage note in the record
Efficiency	Proportion of visits completed without repeat diagnostic studies
User-centeredness	Net promoter score segmented by preferred language

Stories from the Field: Cultural Pillar Spotlight[6]

Boston Children's Hospital offers a powerful case study in how virtual ICU models can extend high-acuity care across borders without compromising quality. Through a real-time digital collaboration with tertiary centers in India, Boston specialists provide pediatric critical care oversight using secure tele-ICU platforms. The program bridges **Physical Pillar** access gaps in remote or under-resourced regions, while simultaneously elevating clinical effectiveness by embedding U.S. protocols, escalation algorithms, and coaching for local teams.

Unlike many telehealth ventures that scale quickly but shallowly, Boston's model is designed for depth. The virtual ICU rounds do not replace local care; they enhance it through **Trust/Knowledge Pillar** access, using structured decision support, shared care plans, and co-managed escalation pathways. The result is not just better throughput or connection rates, but measurable improvements in stabilization times and transfer avoidance.

Quality here is not theoretical; it is delivered in the form of lives saved and ICU stays shortened. And it is sustained through cultural humility. Training content is localized, clinician workflows are adapted, and Boston teams operate as consultants, not commanders. This balance of **Cultural Pillar** access and **Digital Pillar** access shows what "global without imperial" can look like in digital health. Boston Children's does not just export expertise; they co-produce quality at a distance, making this a model for what digital equity and effectiveness can look like in practice.

India's eSanjeevani platform has facilitated over 340 million digital consultations, making it the world's largest telemedicine program.[7] But its real innovation is architectural: by combining provider-to-patient visits with doctor-to-doctor consults, it embeds clinical escalation and team-based care into the workflow. This strengthens clinical effectiveness and safety, particularly in underserved rural regions where access is limited **(Physical Barrier)**. The platform is also free, removing a major **Financial Barrier**. Its integration with regional centers of excellence further underscores that scaling quality requires protocols, not just portals.

[6] Boston Children's Hospital. Partner organizations. Boston Children's Hospital. Accessed May 28, 2025. `https://www.childrenshospital.org/international/organizations`.

[7] Economic Times. Over 34 crore patients provided consultation through eSanjeevani. Accessed May 28, 2025. `https://m.economictimes.com/industry/healthcare/biotech/healthcare/over-34-crore-patients-provided-consultation-through-esanjeevani-platform-rajya-sabha-told/articleshow/118149172.cms`.

Stories from the Field: Trust/Knowledge Barrier Spotlight[8]

Dr. Nneka Sederstrom calls quality "hollow" when structural racism blocks the path to care. Minnesota's redlined geography still shapes healthcare deserts. Fancy digital pods may identify disease but, "now you do not have a facility to go to; you do not have a surgeon," she explains. Rural Native and Hispanic residents face the same broadband gaps as white farmers, but worse outcomes, because, according to Dr. Sederstrom, "systemic racism; lack of insurance; low-income opportunities; and an environment where your cortisol never relaxes" persist.

She recounts a Chicago colleague who built a trauma outpost midway between the South Side streets and a distant academic center. The station cut ambulance travel from eight to three and a half minutes and lives were saved, yet philanthropy ran out and hospitals judged the project "not a good return on investment." For Sederstrom, that decision proves spreadsheets alone cannot define quality.

Bias also rides digital rails. EHRs that auto-populate differential diagnoses lists from strict demographic race info are at risk of what she calls "auto racism." Large language models may amplify the same flaw because, as Dr. Sederstrom says, "I have no idea who trained ChatGPT not to be racist; all of us are inherently racist." True quality, she insists, starts with dismantling structural barriers. Technology and metrics can confirm progress only after that first step. Until housing patterns, transport links, and algorithmic training data are reengineered, the label *high quality* remains aspirational rather than factual.

Managing Risk: Protecting Patients and Providers

Risk in digital care stretches far beyond stray software bugs. It now spans biased machine learning models, ransomware that paralyzes revenue cycles, diagnostic engines that overfit on narrow data, and patient records that splinter as people move between portals. The 2024 Change Healthcare ransomware incident shut down claims flows for weeks, exposed millions of files, and prompted Congress

---

[8] Sederstrom NO. Interview with Matt S. March 31, 2025.

to draft tougher breach-reporting rules, a reminder that connectivity without cyber-resilience is a false bargain.[9]

Other nations are piloting context-specific models instead of copying a single template. The United Kingdom's Medicines and Healthcare Products Regulatory Agency launched the AI Airlock, a sandbox where developers, clinicians, and ethicists co-test high-risk tools inside a controlled clinical setting before nationwide release.[10] Singapore's Health Sciences Authority is drafting a limited exemption that lets public hospitals deploy home-grown diagnostic algorithms inside a monitored sandbox while data on safety, equity, and efficacy accumulate.[11] Both programs emphasize iterative evidence generation, not one-and-done approval.

No rulebook, however elegant, can foresee every hazard. Frameworks such as the National Institute of Standards and Technology AI Risk Management Framework urge boards to treat algorithmic bias, misinformation, and cyber threats as enterprise-level risks that demand continuous measurement and public transparency.[12] In practice, that means:

- Tying executive bonuses to rolling safety-and-inclusion scorecards
- Publishing model cards that disclose data lineage and known blind spots
- Funding red-team audits that probe for demographic or clinical drift every quarter

Governance in the digital age therefore becomes a layered defense, not a static barrier: adaptive regulation at the national level, sandbox testing at the system level, and risk-informed culture at the board level. Companies that embed these layers early are less likely to repeat lessons of the past and more likely to earn durable trust from patients, clinicians, and payers.

---

[9] Minemyer P. One year later: lessons learned from the Change Healthcare cyberattack. Fierce Healthcare. Accessed May 27, 2025. https://www.fiercehealthcare.com/health-tech/one-year-later-lessons-learned-change-healthcare-cyberattack.

[10] Medicines and Healthcare Products Regulatory Agency. AI Airlock: the regulatory sandbox for AIaMD. Accessed May 27, 2025. https://www.gov.uk/government/collections/ai-airlock-the-regulatory-sandbox-for-aiamd.

[11] Health Sciences Authority Singapore. Public consultation on the proposed exemption for AI software as a medical device. Accessed May 27, 2025. https://www.hsa.gov.sg/announcements/public-consultation/proposed-exemption-AI-SaMD.

[12] National Institute of Standards and Technology. *AI Risk Management Framework*. Accessed May 27, 2025. https://www.nist.gov/itl/ai-risk-management-framework.

Next Shift Quick Wins: Five Board Questions That Instantly Expose Weak AI Risk Governance

- Where in your AI lifecycle do humans have explicit veto power, and how is that documented?
- Which executive is personally accountable for model-drift monitoring and reporting to the board?
- Can you trace every production model back to a signed-off risk assessment covering bias, security, and failure modes?
- What is your mean time to respond (MTTR) to an AI incident that harms patients or materially affects operations, and when was it drilled last?
- How are you auditing third-party AI vendors for updates, cybersecurity patches, and compliance with your own governance policy?

Next Shift Quick Wins

ELEMENT	WHY IT MATTERS
Real-world performance monitoring	Avoids overreliance on lab-validated accuracy
Adaptive risk frameworks	Keeps pace with tech evolution
Clinician-in-the-loop design	Prevents automation bias

Psychological Safety as a Core Dimension of Quality

New York City Health Commissioner Dr. Dave Chokshi looked back on the first pandemic wave and named a second pathogen: misinformation. "The degree to which misinformation is able to take root is deeply concerning to me," he told CBS News during his 2022 exit interview. He tied that risk to racism and other structural inequities that already erode trust.[13] Chokshi's warning reframes digital quality.

---

[13] Moore J. Health commissioner Dave Chokshi reflects on leading NYC through the COVID pandemic. *CBS News New York*. Published March 9, 2022. Accessed May 27, 2025. https://www.cbsnews.com/newyork/news/nyc-health-commissioner-dr-dave-chokshi-covid-pandemic-look-back/.

Accuracy and uptime are necessary, yet they collapse if the public does not believe a platform is worthy of trust.

Quality programs must therefore score tools on "felt safety" as well as accuracy. Metrics might include portal dropout rates after a cyber event, misinformation velocity on local social channels, and community sentiment surveys that reveal whether risk explanations are understood. Boards that tie executive bonuses to both technical uptime and psychological trust signal that these guardrails are inseparable parts of the same quality framework.

Weekly livestreams during the New York surge paired epidemiologic charts with community leaders who translated guidance into everyday choices. That communication strategy did not replace encryption or audit logs, it sat beside them, turning technical quality into lived quality. Future digital rollouts should begin with the same dual checklist: prove the code is safe, then prove people know it is safe. Anything less repeats the errors that destabilized Change Healthcare.[14]

Measuring Success: What Really Matters in Digital Health?

Hospital quality dashboards still spotlight mortality, infection, and 30-day readmission. These numbers remain vital, yet they miss the signal that decides whether a digital tool sinks or scales. What's the right denominator for a chatbot? Or a no-show-prediction algorithm? A fall prediction algorithm that pings too often will be silenced no matter how many lives it might save. A chatbot that resolves 90 percent of inquiries but angers the remaining 10 percent will tank a health plan's NPS (Net Promoter Score) and trigger churn. Funding trends confirm the shift. Rock Health analysts note that provider and employer buyers now set deliberate thresholds for return on investment and clinical effectiveness for current and prospective vendors and that reimbursement will tighten for offerings that cannot prove their clinical value.[15]

So how do you choose the right metrics? A metric begins with the right denominator. For patient monitoring, the unit is not discharges; it is minutes from alert to intervention. A multicenter study of 905 ward patients found a median lead time of at least 18 hours between

---

[14] No statistically significant cupcakes were harmed in the making of this chapter; however, several sprinkles were rearranged in the name of quality improvement.

[15] Sussman A, Simmons D, Kaganoff S, Chamberlain R. 2023 look ahead: boring is the new black. *Rock Health.* Published December 5, 2022. Accessed May 27, 2025. https://rockhealth.com/insights/2023-look-ahead-boring-is-the-new-black/.

the first remote-generated warning and documented deterioration, giving staff a full shift to act.[16] For an emergency department triage model, the denominator is decision minutes, GPT4 trimmed that interval and matched or exceeded resident diagnostic accuracy in 100 consecutive cases.[17] Diagram 9.4 shows a comparison of traditional healthcare metrics with an example use case in the "digital age."

TRADITIONAL CLINICAL METRIC	DIGITAL-AGE METRIC (EXAMPLE USE CASE)
30-day all-cause readmission rate: Percent of discharges that bounce back within 30 days; anchor of the CMS Hospital Readmissions Reduction Program[18]	**Alert-to-intervention time in remote patient monitoring (RPM):** Median minutes from a vital-sign alert to clinical action; for example, an early-warning RPM platform gave staff ≥18 hours of extra lead time before deterioration
Door-to-balloon (D2B) time for STEMI: Goal <90 min from ED arrival to coronary reperfusion; long-standing ACC/AHA benchmark[19]	**AI-triage "time-saved" index:** Minutes shaved off first therapeutic move when machine-learning triage ranks cases
Clinic no-show rate for in-person visits: Share of scheduled encounters missed without notice; tracked by MGMA and payers	**Virtual-visit completion rate:** Percent of scheduled telehealth appointments successfully connected

Diagram 9-4: Metrics That Matter.

[16] Lakshman P, Gopal PT, Khurdi S. Effectiveness of remote patient monitoring equipped with an early warning system in tertiary-care hospital wards. *J Med Internet Res*. 2025;27:e56463. Accessed May 27, 2025. https://www.jmir.org/2025/1/e56463.

[17] Hoppe JM, Auer MK, Strüven A, Massberg S, Stremmel C. ChatGPT with GPT-4 outperforms emergency department physicians in diagnostic accuracy. *J Med Internet Res*. 2024;26:e56110. Accessed May 27, 2025. https://www.jmir.org/2024/1/e56110.

[18] Centers for Medicare & Medicaid Services. Hospital-wide 30-day all-cause unplanned readmission measure specifications. Published 2024. Accessed May 27, 2025. https://qpp.cms.gov/resources/document/19e89489-50dd-42c3-b363-281cc4c4c557.

[19] Zeng X, Chen L, Chandra A, Zhao L, Ma G, Roldan FJ, Wei H, Pan W, Li W. Narrative review: updates and strategies for reducing door-to-balloon time in ST-elevation myocardial infarction care. *Front Cardiovasc Med*. 2025 Mar 31;12:1509365. doi:10.3389/fcvm.2025.1509365. PMID: 40231025; PMCID: PMC11994590.

Worked Example: Turning 15 Minutes into Margin

So how can a second here or a minute there really add up? Imagine that you are caring for sepsis patients in the ICU. These patients can have complex needs, are hooked up to *all* the monitors, and can deteriorate quickly. What if you can alter the staff to start antibiotics 15 minutes sooner than they had been last quarter?

That seemingly modest 15-minute head start isn't just a trophy for the quality dashboard, it's a hard-nosed business lever. Clinically, shaving even a quarter-hour off "alert-to-antibiotic" lets you interrupt the sepsis spiral sooner, lopping nearly double that time off length-of-stay (LOS) per patient.

Operationally, those reclaimed hours aggregate into 25 extra bed-days a year in a 1,000-case service line, enough slack to absorb a holiday surge without hallway boarding. That translates to roughly $78,000 in estimated variable cost avoidance and another $13,000 in pure contribution margin. This is cash you can reinvest in more AI tooling or, dare I say, better coffee for the rapid-response team (tacos would be even better!).

In other words, a quarter-hour nudge on the stopwatch becomes almost a six-figure nudge on the bottom line. This is proof that every quality metric should be walked all the way across the bridge to LOS and dollars before it lands in the board deck (see Diagram 9.5).

STEP	MATH	RESULT
1. **Translate the 15-min gain into LOS impact.**	Multicenter data shows that each one-hour treatment delay adds 0.10 hospital days (≈ 2.4 h) to a sepsis stay[a,20]	ΔLOS = −0.025 days/patient
	So a 0.25 h (15-min) improvement shaves: 0.25 h $\times$ 0.10 d/h = 0.025 days (≈ 36 min) per patient	
2. **Annualize the bed-days freed.**	1,000 sepsis admissions/ yr $\times$ 0.025 d = 25 bed-days opened up	25 bed-days/yr

(continues)

[20] Zhang D, Micek ST, Kollef MH. Time to appropriate antibiotic therapy is an independent determinant of post-infection ICU and hospital lengths of stay in patients with sepsis. *Critical Care Medicine*. 2015;43(10):2133–2140. Accessed June 1, 2025. https://www.jvsmedicscorner.com/ICU-Inflammation_and_infection_files/Appropriate%20Antibiotic%20Treatment%20in%20Severe%20Sepsis%20and%20Septic%20Shock%20Timing%20is%20everything.pdf.

(Continued)

STEP	MATH	RESULT
3. **Convert bed-days to variable cost saved.**	Avg variable cost of a med-surg day $3,132[b,21] 25 d × $3132 = $78,300 freed capacity cost	~$78,000/yr
4. **Capture the margin opportunity.**	If your contribution margin is $532 per bed-day (17% on $3132): 25 d × $532 = $13,300 potential incremental margin (or capacity to treat more patients)	~$13,000/yr
5. **Summarize for the board.**	A mere quarter-hour faster rescue in sepsis unlocks 25 bed-days and $91,600 in combined cost avoidance + margin expansion each year, at zero added head count	~$92,000 headline

[a]Assumption: Increase in LOS of about 0.1 days per one-hour delay.
[b]Assumption: Avg inpatient cost per day is $3,132 (varies based on state and hospital structure).

Diagram 9-5: The Math Behind the Measure. Your Sepsis Early-Warning Algorithm Now Pages the Rapid-Response Team 15 Minutes Sooner Than Last Quarter.

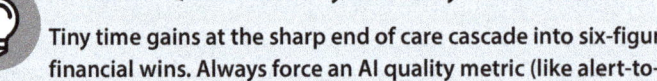

Next Shift Quick Wins: Key Takeaway for Directors

Tiny time gains at the sharp end of care cascade into six-figure financial wins. Always force an AI quality metric (like alert-to-intervention) to cross the bridge into LOS (length of stay) and dollars so stewardship stays front-of-mind.

(Swap in your own hospital's LOS elasticity, case volume, and cost figures for a custom calculation.)

Balanced Score Card?

We've talked about metrics, and every system and organizations has their favorites. But we all know every fresh MBA loves a "balanced score card." What does this mean for "scoring" quality and access in today's everchanging environment? Instead of one-size-fits-all KPIs, successful programs use multidimensional metrics: from

[21] Kaiser Family Foundation. Hospital expenses per adjusted inpatient day. KFF State Health Facts. Accessed June 1, 2025. https://www.kff.org/health-costs/state-indicator/expenses-per-inpatient-day/.

patient retention and NPS, to operational efficiency and clinician workload reduction.

Successful programs layer four metric families instead of a single return on investment line:

- **Clinical Effectiveness:** Alert-to-intervention time; false-negative rate; patient level outcomes such as escalations avoided.

- **Experience and Loyalty:** Net Promoter Score (NPS); patient retention at 90 days; dropout curves for chatbots and asynchronous modules.

- **Operational Efficiency:** Clicks removed; messages deflected; minutes returned to the schedule.

- **Financial Sustainability:** Cost per interaction; margin per episode; contract renewal rate.

If you think a level up about how the executive team can be held accountable for new metrics like these, there are a variety of options. Tying bonuses to this balanced scorecard aligns incentives across functions. The CFO cares about margin, the CNO cares about clicks avoided, the Chief Experience Officer watches NPS and social sentiment; the CMO examines false negatives and deterioration lead time. The list goes on, and it truly becomes an entire executive team that is balanced in their goals and is incentivized together to do so. When the CFO, CNO, and CXO walk into a bar—sorry, boardroom—and all leave smiling, you know the scorecard's working.

Digital systems can widen access gaps as easily as they close them.

(Yes, that's me shouting from the soapbox again!)

To ensure these metrics are working to improve healthcare access, every metric should be stratified by language, race, and income. Digital systems can widen gaps as easily as they close them. If you've read to Chapter 9 and haven't really absorbed that, I'll restate again: Digital systems can widen access gaps as easily as they close them. The same telehealth study that halved missed appointments might also uncover persistent disparities for households without broadband. Continuous stratification turns quality from a press release into a numeric target, enabling course correction before harm accrues.

Digital tools are ever evolving, and we'll need to evolve metrics to go with them. Borrowing from the FDA Total Product Lifecycle model, organizations now publish live dashboards that display accuracy drift, override frequency, and uptime. When drift crosses a preset threshold, the algorithm downgrades to advisory mode until retrained data restore performance. This practice treats safety and efficacy as dynamic properties rather than one and done checkpoints.

So, here's the takeaway: classic metrics still reveal outcomes, yet digital care layers new "speed-to-action" and "completion-of-connection" indicators that better capture quality in remote monitoring, algorithmic triage, and virtual visits. This doesn't mean you get rid of the traditional metrics; it means you use the right metrics for the right use cases. It also means that you need to get smarter about what you are measuring. As the adage goes, "we manage what we measure."

Building Standards: Creating Frameworks for the Future

Without interoperability and shared standards, digital care becomes patchwork care. The ONC's Trusted Exchange Framework in the United States is a step forward. But most digital systems still rely on vendor-specific solutions. In Estonia and Denmark, standards have essentially become the backbone of digital quality.

In most countries, while digital health innovation is fast; interoperability is an afterthought. Estonia and Denmark prove the reverse can be true, and more sustainable. By embedding national standards for identity, data sharing, and record access into their healthcare systems, these countries demonstrate that quality in digital care is inseparable from infrastructure.[22] Estonia's X-Road and Denmark's MitID systems allow secure, standardized health data to move across clinics, pharmacies, hospitals, and insurers. These are not just technical conveniences; they are foundational quality enablers. When every clinician accesses the same verified record in real time, you eliminate delays, omissions, and duplication. That's not just efficiency, it's safety. That's not just access, it's inclusion. In a fragmented system, those same transitions of care become high-risk events.

Standards also shape what's measurable. In Denmark, unified APIs allow health outcomes and care quality to be tracked nationally in near-real time.[23] Antibiotic stewardship dashboards, readmission

---

[22] Ministry of Economic Affairs and Communications, Republic of Estonia. *X-Road: foundation of e-Estonia*. Published 2024. Accessed June 1, 2025. https://e-estonia.com/solutions/interoperability-services/x-road/.
[23] Danish Health Data Authority. *National APIs for health-quality monitoring: annual status report 2024*. Published 2024. Accessed June 1, 2025. https://sundhedsda tastyrelsen.dk/-/media/sds/filer/english/api_quality_report.pdf.

audits, and maternal health interventions can all be assessed with granularity because the data is interoperable by design. This drives clinical effectiveness across the board; not by micromanaging clinicians, but by giving them tools that speak the same language and by decreasing **Digital Barriers**.

Stories from the Field: Digital Pillar Spotlight

Launched in June 2023, the GDHCN is the World Health Organization's open-source trust network for digitally signed health credentials.[24] It builds on the EU Digital COVID Certificate rails. Participating countries publish public keys through a shared gateway so any border, clinic, or pharmacy can verify that a QR code or smart card is genuine without ever touching the underlying patient data.[25] By late 2024 more than 80 member states were live on the network,[26] and the 78th World Health Assembly extended the Global Strategy on Digital Health through 2027 while mandating the next phase for 2028–2033 (see Diagram 9.6).[27]

Why boards should track the GDHCN:

- ▪ **One Standard, Many Use Cases:** COVID-19 vaccination cards were only the proof-of-concept. The same rails now carry yellow-fever proof-of-vaccination, polio laboratory results, cross border e-prescriptions, and international patient summaries.[28]

---

[24] World Health Organization. *Global Digital Health Certification Network Overview.* Accessed May 31, 2025. https://www.who.int/initiatives/global-digital-health-certification-network.

[25] European Commission. *EU Digital COVID Certificate becomes the first building block of the GDHCN.* Accessed May 31, 2025. https://commission.europa.eu/strategy-and-policy/coronavirus-response/safe-covid-19-vaccines-europeans/eu-digital-covid-certificate_en.

[26] World Health Organization. *WHO global network expands digital health certification for Hajj pilgrims.* Accessed May 31, 2025. https://www.who.int/news/item/21-10-2024-who-global-network-expands-digital-health-certification-for-hajj-pilgrims.

[27] World Health Organization. *World Health Assembly endorses the extension of the Global Strategy on Digital Health to 2027.* Accessed May 31, 2025. https://www.who.int/news/item/23-05-2025-world-health-assembly-endorses-extension-of-the-global-digital-health-strategy-to-2027.

[28] European Commission. *EU Digital COVID Certificate becomes the first building block of the GDHCN.* Accessed May 31, 2025. https://commission.europa.eu/strategy-and-policy/coronavirus-response/safe-covid-19-vaccines-europeans/eu-digital-covid-certificate_en.

- **Privacy by Design, No Central Database Moves:** Only public keys transit the gateway, which keeps the network friendly to GDPR, HIPAA, and Kenya's Data Protection Act.

- **Rapid Uptake, Early Mover Advantage:** Eighty-plus countries already issue globally verifiable credentials. Alignment now positions providers for cross-border tele-ICU billing and other digital services when they mature.

Stories from the Field: Digital Pillar Spotlight

Fatima is a 32-year-old elementary school teacher in Jakarta. Two months before leaving for Hajj, she opened the *Satu Sehat* app on her phone and requested the new electronic yellow fever certificate required for travelers from Indonesia.[29] The Ministry of Health automatically pulled her immunization record, packaged it as a FHIR bundle, signed that bundle with its private key, and published only the matching public key to the WHO Global Digital Health Certification Network (GDHCN) gateway. Fatima saw a single QR code appear onscreen and saved a wallet copy for offline use. No personal data ever left Indonesia.

Departure day at Soekarno Hatta Airport moved quickly. Check-in staff scanned the code. Their airline app validated the signature against the GDHCN key list and printed *VERIFIED* on her boarding pass. Nine hours later she landed in Jeddah, where a border officer tapped the same code. The Saudi *Sehat* verifier reached the GDHCN gateway, confirmed the Indonesian signature in 80 milliseconds, and the gate that opened. Fatima's phone never connected to a foreign server, yet everyone trusted the result.

Midway through the pilgrimage Fatima slipped on the marble during tawaf and strained her ankle. At the Mecca urgent care clinic, the triage nurse scanned the identical QR code. The EHR ingested her vaccination bundle straight into the encounter note so the physician could rule out yellow fever and focus on the sprain. The visit generated a digital prescription for ibuprofen that also rode the GDHCN rails (see Diagram 9.6). Fatima picked up the medicine at a nearby pharmacy without filling out a single paper form.

---

[29] World Health Organization. *WHO global network expands digital health certification for Hajj pilgrims.* Accessed May 31, 2025. https://www.who.int/news/item/21-10-2024-who-global-network-expands-digital-health-certification-for-hajj-pilgrims.

When she returned to Jakarta, her family doctor opened her file and saw the pharmacy dispense record already reconciled. The same globally trusted credential followed her from home to airport to clinic to pharmacy and back again, keeping care moving and paperwork invisible.

PILLAR	STRATEGIC IMPACT	ILLUSTRATIVE EXAMPLE
Physical	Faster clearance at borders; quicker ED triage	The Hajj Health Card pilot improved clearance for 250,000 international patients[30]
Digital	Truly interoperable FHIR/IPS payloads	One QR payload flows into an EHR or a travel app without rekeying
Financial	Lower administrative costs	The EU saved billions in paper verification during the Digital COVID Certificate roll-out
Cultural	Multilingual certificate templates	Automatic localization ends "lost in translation" yellow cards
Trust/ Knowledge	Global cryptographic proof chain	Counterfeit vaccine certificates and misinformation fall dramatically

Diagram 9-6: The Impact of GDHCN on the Pillars.

Board-level "watch-outs":

- **Vendor Alignment:** Insist that any new patient-facing app or remote monitoring platform used in member states can export GDHCN-ready FHIR bundles.

- **Governance Overlap:** Information-security committees should own key-rotation schedules, while digital-quality teams track false positives and negatives during credential scans.

- **Future-Proofing:** The WHO has signaled that the network will anchor the 2028–2033 digital health strategy. Early alignment creates first-mover leverage when cross-border billing arrives.

---

[30] European Commission. *EU Digital COVID Certificate becomes the first building block of the GDHCN.* Accessed May 31, 2025. https://commission.europa.eu/strategy-and-policy/coronavirus-response/safe-covid-19-vaccines-europeans/eu-digital-covid-certificate_en.

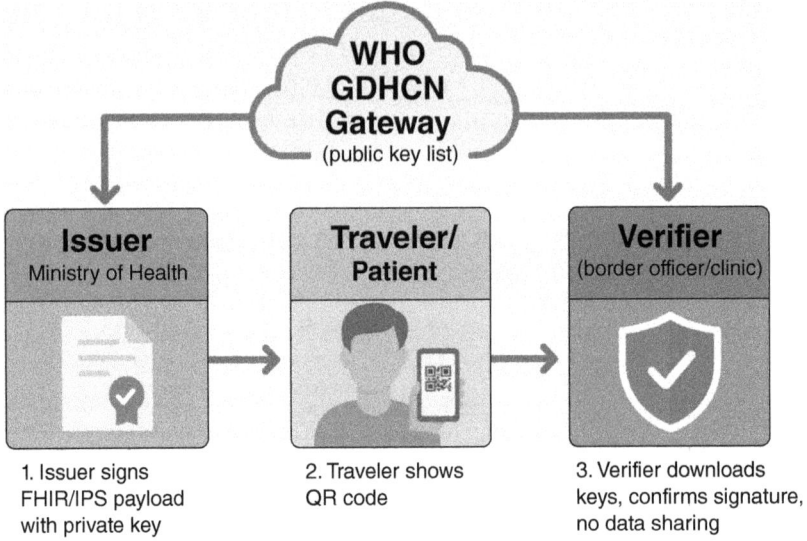

1. Issuer signs FHIR/IPS payload with private key
2. Traveler shows QR code
3. Verifier downloads keys, confirms signature, no data sharing

Diagram 9-7: Data Flow of the WHO GDHCN.

Sound-bite for your next board deck: "Think of the GDHCN as HTTPS for health data; one globally trusted lock icon lets care follow the patient wherever the patient goes."

… now for the rest of us … HTTPS is like sealing your online letters in a locked envelope. It scrambles everything you send and receive so snoops only see gibberish, and it double-checks that the site you are talking to is really the site you meant to visit (see Diagram 9.7).

 Next Shift Quick Wins

What makes a standard quality-enabling? Use this checklist to evaluate whether a digital health standard truly advances care quality:

■ **Interoperability:** Can data move seamlessly across providers, systems, and platforms?

■ **User Transparency:** Can patients see who accessed their records and for what purpose?

■ **Scalability:** Can new tools plug into the system without rewriting infrastructure?

■ **Security and Access Control:** Is sensitive data protected while being practically usable in clinical workflows?

- **Auditability:** Can performance (e.g., access delays, errors) be measured and traced?

- **Cultural Flexibility:** Does it support multilingual content and localization for diverse populations?

- **Built for Metrics:** Does the structure allow for real-time tracking of quality indicators?

- **Future-Proofing:** Is the standard extensible for AI, wearables, and telehealth without losing integrity?

Ultimately, patients like Fatima benefit from standards. In her story, both countries gave people real-time transparency into who accessed their health records and why, using citizen-facing interfaces that reinforce the **Trust/Knowledge Pillar**. These aren't token features, they're structural signals that quality is co-produced, not extracted. Users aren't just passive recipients of care; they're informed participants in it. And critically, this isn't built on vendor-specific shortcuts or startup APIs. These are national standards. They are enforced by law, designed to evolve, and made available as public goods. That's what makes innovation possible without introducing chaos. New AI tools or virtual care platforms don't have to reinvent interoperability; they plug into a system that was designed for safety, transparency, and scalability.

In contrast to the United States, where digital health often operates in "walled gardens" and piecemeal pilots, Estonia, Denmark, and the WHO GDHCN show what's possible when the rules of the road are clear, enforced, and centered on outcomes. Without shared standards, digital care is fragile. With them, it becomes resilient and far more likely to deliver quality at scale.

The Role of Government in Digital Health Quality

Governments are the single most important accelerators or barriers to quality digital care. Regulatory frameworks and policy decisions heavily influence whether digital health solutions become mainstream or remain on the periphery. Before COVID-19, rigid telemedicine regulations, including licensing barriers, privacy concerns, and minimal reimbursement, slowed adoption in many countries. During the pandemic, emergency regulations triggered a rapid expansion of telehealth worldwide, highlighting the power of more flexible policies to widen access safely.

Fact Check: Digital Policy

1. India's digital health mission has issued over 400 million health IDs, enabling unified patient records across states.[31]

2. Australia's opt-out "My Health Record" held digital files for more than 23 million residents in 2019, and has continued to grow, giving any clinician instant access.[32]

3. Brazil made telemedicine permanent in 2022; the new law obligates the public SUS system to reimburse virtual visits.[33]

The next challenge is to make these temporary measures permanent while addressing data security, interoperability, and cross-border telehealth concerns. There is a general lack of transnational regulatory standards, biases against technology-based medicine, and limited interoperability. Creating coherent, cross-border legal frameworks could spark further innovation and help telehealth transcend regional limitations (see Diagram 9.8).

Beyond telemedicine, AI-driven healthcare adds another layer of regulatory complexity. Ensuring clinical validation, fairness, and patient safety for AI-powered systems remains a central policy issue. Regulatory bodies, like the FDA in the United States, are already drafting guidelines for AI-based medical devices. Effective data governance is also vital: patient privacy must be protected, but beneficial data sharing for clinical research and population health improvement should not be unduly hampered.

---

[31] Murthy N, Chandrasekaran A, Bane S, et al. Effects of an mHealth voice message service (mMitra) on maternal health practices in urban India. *BMC Public Health*. 2020;20:820. Accessed May 31, 2025. https://bmcpublichealth. biomedcentral.com/articles/10.1186/s12889-020-08965-2.

[32] Babylon Holdings Ltd. Babylon launches AI-powered triage tool in Rwanda to support national digital health strategy [press release]. Published December 3, 2021. Accessed May 31, 2025. https://www.businesswire.com/news/home/20211203005293/en.

[33] Ada Health. World's first AI health guidance app in Swahili launched by Ada Health [press release]. Published November 19, 2019. Accessed May 31, 2025. https://about.ada.com/press/191119-worlds-first-ai-health-guidance-app-in-swahili/.

ACCESS PILLAR	GOVERNMENT REGULATORY LAVER
Physical	Class II/III device clearance (510(k))
Digital	Rural broadband mandates and spectrum subsidies
Financial	Telehealth payment parity and value-based models
Cultural	Official language, localization requirements
Trust/Knowledge	Data ethics, privacy and security laws

Diagram 9-8: Government Levers Across the Access Pillars.

Iterative Approaches: Adapting to Real-World Feedback

Early digital health pilots often stumbled on the simplest barrier: language. (It happens today too!) This prevents a quality experience and a quality product. In Kenya, the first cervical-screening apps shipped only in English. Community health volunteers froze at untranslated terms, and screens stalled whenever the cellular signal dipped. Uptake plateaued until developers doubled back, translated every prompt into Kiswahili and added an offline cache. The pattern has replayed across several continents, and each reboot offers the same lesson—localization plus offline resilience turns cultural friction into quality scale. I saw this myself at multiple organizations, trying to bring in multiple platforms post-acquisition. We wanted to scale the platform and hit international markets. However, each country had their own rules and local regulations. Additionally, the language of medicine and the language of patients is often different. So even if your EHR or application suits physicians and nurses in one language, the portal or outreach tools may need to be geared for patients in several languages, and may need to be at different reading levels.

Take India's mMitra voice-message program. Every maternal-health tip is played in Hindi and Marathi. All 145 messages sent to patients were recorded in both languages. In a pseudo-randomized trial of 2016 women, of those using the fully localized version, over 60 percent of women reported that mMitra had improved their health awareness, and that they had shared the information with family members.[34]

Rwanda then moved the idea into artificial intelligence. In late 2021, Babyl, a partnership with Babylon Health, unveiled an AI triage tool fully localized for Rwanda and accounted for local language, epidemiology, culture, and health system pathways. More than 2 million patients were registered and ran through the system in the first year alone.[35] Even commercial apps have learned the rule. Ada Health launched a Kiswahili symptom checker in 2019, claiming it "unlocked access for more than 100 million people in East Africa," and highlighted language work as the price of entry.[36]

Across these stories, localization raises the **Cultural Pillar** by speaking users' own words, while offline-first design reinforces the **Trust/Knowledge Pillar** by letting frontline workers counsel straight from the screen. Access is not just coverage, it is comprehension. True quality in digital health means iterating everywhere except the operating theatre. Iteration is not failure, it is respect.

So, for every product manager, TPM, or program lead, the mandate is clear—bake a local language milestone into sprint one, cache data on the device before the first field test, and spend a shadow day each quarter watching real users highlight the next rough edge to sand away.

What Separates Success from Failure?

The pandemic-fueled surge in telemedicine brought healthcare's digital frontier to the mainstream, accelerating adoption at a pace once unimaginable. From homes to hotel rooms, patients connected with clinicians over Wi-Fi, and venture capital poured in. But scale alone has never been the true measure of success in healthcare.

---

[34] U.S.: The rise and fall of early telehealth startups (e.g., Babylon Health, Forward). Sifted. Accessed February 7, 2025. https://sifted.eu/articles/teleme dicine-startups-problems.

[35] The fall of Babylon: Failed telehealth startup once valued at nearly $2B goes bankrupt and sold for parts. TechCrunch. Published August 31, 2023. Accessed February 7, 2025. https://techcrunch.com/2023/08/31/the-fall- of-babylon-failed-tele-health-startup-once-valued-at- nearly-2b-goes-bankrupt-and-sold-for-parts/.

[36] Does Forward Health's failure mark the winter of telehealth? ICT & Health. Accessed February 7, 2025. https://ictandhealth.com/news/does-for ward-healths-failure-mark-the-winter-of-telehealth.

The post-pandemic years revealed a sobering truth: quality, not novelty, is the defining metric of survival.

This is not a new lesson. Even during NASA's STARPAHC project in the 1970s, technologists dreamed of commercializing telemedicine. Yet the same stumbling blocks persist—an inability to demonstrate long-term value, tenuous reimbursement, and poorly aligned incentives. Digital health may be borderless, but it is not immune to the same foundational flaws that plague brick-and-mortar systems.

Take Babylon Health, once a darling of the telehealth scene and valued at nearly $2 billion.[37] Fueled by deep-pocketed investors and bold ambitions, Babylon rapidly expanded across the UK and United States. But cracks appeared beneath the glossy surface. By 2021, safety concerns around its chatbot triage system led to regulatory scrutiny and public distrust. In 2023, a failed acquisition by MindMaze ended in bankruptcy stateside and delisting from the New York Stock Exchange.[38] Babylon's UK operations were acquired by eMed, but the implosion of its U.S. business sent a clear message: scale without reliability is not scale worth building.

Another cautionary tale is Forward Health, established in 2016 with ambitions to revolutionize primary care. Its AI-powered "Forward Care Pods," placed in locations from malls to gyms, promised convenience and futuristic technology. Yet the costly, high-tech kiosks and steep monthly fees (for what was essentially primary care) failed to generate enough demand. Forward Health shut down in fall 2024, underscoring the importance of blending user-centric design with financial sustainability.[39] In the end, despite their technological flair, Forward's pods did not offer patients, or investors, a compelling reason to buy in. It's a bit like opening a fancy new restaurant with no staff, no kitchen, and no recipes; sure, the concept might be brilliant, but nobody's getting fed.

Statistics highlight the steep climb most telehealth startups face. Roughly 21 percent fail in their first year, 30 percent by their second,

---

[37] Does Forward Health's failure mark the winter of telehealth? ICT & Health. Accessed February 7, 2025 `https://ictandhealth.com/news/does-forward-healths-failure-mark-the-winter-of-telehealth`.

[38] The fall of Babylon: Failed telehealth startup once valued at nearly $2B goes bankrupt and sold for parts. TechCrunch. Published August 31, 2023. Accessed February 7, 2025. `https://techcrunch.com/2023/08/31/the-fall-of-babylon-failed-tele-health-startup-once-valued-at-nearly-2b-goes-bankrupt-and-sold-for-parts/`.

[39] Does Forward Health's failure mark the winter of telehealth? ICT & Health. Accessed February 7, 2025. `https://ictandhealth.com/news/does-forward-healths-failure-mark-the-winter-of-telehealth`.

50 percent by their fifth, and 70 percent by their tenth.[40] One major stumbling block is the lack of proper credentialing and reimbursement. If government agencies, such as the Centers for Medicare & Medicaid Services (CMS), or private insurers do not cover a telehealth service, organizations are unlikely to adopt it. Gaining traction, therefore, requires time, trusted partnerships with regulators, and strategic approaches that demonstrate real value to payers and patients alike.

Telehealth startups: Roughly 21 percent fail in their first year, 30 percent by their second, 50 percent by their fifth, and 70 percent by their tenth

These cautionary tales make one thing clear: in digital healthcare, quality is not a feature, it's the foundation. Success comes not from having the most elegant codebase or sleekest interface, but from building systems that are credible, clinically sound, and integrated into patients' real lives. That means earning regulatory trust, embedding in reimbursement frameworks, and prioritizing user experience from the first click to the final outcome. Companies that thrive in this space understand that digital access must be paired with **Cultural Pillar** humility, **Financial Pillar** feasibility, and seamless interoperability. They do not just measure clicks or downloads; they measure patient outcomes, system inclusivity, and sustained use. They recognize that in healthcare, the "minimum viable product" must still meet the maximum expectations of safety, quality, and compassion.

Taken together, these stories illustrate that, while telemedicine can fill critical gaps in healthcare delivery, a novel idea alone does not guarantee success. Thriving in this space demands a patient-focused approach, robust reimbursement strategies, and careful alignment with regulatory frameworks—essentials that many early telehealth companies simply overlooked. Quality in digital health is not about proving what's possible. It's about proving what works. And for whom.

Fact Check

1. Annual venture funding for 2023 closed out at US $10.7B raised across 492 deals, the lowest amount of capital invested in U.S.-based digital health startups since 2019.

2. As of December 31, 2023, at least 17 percent of public digital health companies trading on the NASDAQ or NYSE were noncompliant with

---

[40] Telemedicine startup landscape and trends. Empeek. Accessed February 7, 2025. https://empeek.com/insights/telemedicine-startup/.

listing standards, having traded at or below $1 for over 30 consecutive business days.[41]

3. The Canadian VC ecosystem is more concentrated on seed and later stage investments, whereas the U.S. market distributes its funding more evenly but with a tendency toward larger deals.[42]

Looking Ahead

The healthcare landscape of today scarcely resembles the sweltering house calls I once made in Austin, yet access is still elusive. Technology has lowered some walls and raised others. Telemedicine can pull a specialist onto any screen but broadband gaps and limited digital literacy still keep whole communities offline. Fred with his diabetic foot and Hannah alone on her couch remind us that distance and quality are not the same.

Today's barriers interlock like gears. Distance punishes rural residents, refugees, and people without stable housing or transportation **(Physical Barrier)**. Strain deepens when reimbursement ignores behavioral health and addiction care **(Financial Barrier)**. Cultural identity clashes with systems that do not speak a patient's language or reflect their lived reality **(Cultural Barrier)**. Digital tools reward the connected and the confident. Trust fractures under historic injustice and impersonal care.

These same barriers also map the way forward. Quality in the digital age is adaptability with integrity. A chatbot that lowers missed appointments is useful; one that leaves a patient feeling known and safe is transformative. The best tools thrive in the chaos of daily life; they see social determinants, silent burdens, and personal narratives.

The five pillars of access must be design criteria, not check marks. A truly borderless system erases every invisible wall between a person and the care they deserve; that is how we will measure what matters.

---

[41] Rock Health. *Digital Health Funding Report 2023.* Accessed February 11, 2025. https://rockhealth.com/insights/2023-year-end-digital-health-funding/.

[42] Health Tech Report. Published January 10, 2024. Accessed February 11, 2025. https://ss-usa.s3.amazonaws.com/c/308495893/media/899765a 965eea822a21109278981792/Jan10%20-%20Health%20Tech%20Repo rt%20VF1.1.pdf.

Startup Builder's Box: Quick Hits

QUICK HIT	TO-DO
1. **Run a four metric quality scorecard from day one.**	Pick one clinical, one experience, one operational, and one financial signal and display them on the same dashboard.
2. **Slice every metric by language, race, and income.**	Trigger an alert when any group slips 10% behind the median.
3. **Localize and cache from sprint one.**	Release in the user's language and keep the core workflow working without connectivity.
4. **Time alert to action.**	Convert minutes saved into bed days and dollars for the next board slide.
5. **Court regulators and payers early.**	Schedule an FDA touchpoint or join a payer sandbox this quarter.
6. **Build on open standards.**	Adopt FHIR-compatible interfaces and enroll in a trusted health data exchange before beta launch.

Leaders' End-of-Chapter Action Checklist: Chapter 9: "Quality in the Digital Age"

LEADER	HIGH-IMPACT ACTION TO STRENGTHEN QUALITY/TRUST AND KNOWLEDGE ACCESS
❏ Board Director	Tie 5% of executive bonus pool to a rolling inclusion-and-safety scorecard that includes psychological trust metrics such as patient dropout rate after cyber events and misinformation velocity tracking.
❏ Chief Executive Officer	Charter a cross-functional Digital Quality Task Force to co-develop an inclusion-adjusted NPS and stratified satisfaction index; publish results quarterly.
❏ Chief Information Officer	Add real-time AI model drift, uptime, and override frequency dashboards to the intranet; trigger advisory mode if thresholds are crossed.
❏ Chief Health Information Officer	Mandate clinician transparency protocols in digital settings, including plain-language algorithm disclosures; monitor portal literacy through multilingual completion metrics.
❏ VP Clinical Operations	Require "explainability briefings" for all new digital tools during staff huddles; spot audit decision aid comprehension with patient playback interviews.
❏ VP Nursing & Patient Education	Launch digital trust rounds where nurses review one patient per shift for comprehension, perceived respect, and clarity; log barriers using structured tags.
❏ VP Data & Analytics	Publish model cards for every predictive algorithm used in scheduling or triage; disclose data lineage, training demographics, and performance blind spots.
❏ Telehealth Program Manager	Implement a 30-second "explain-your-tools" script at the start of virtual visits; sample 10 visits per month to verify clarity and completion.
❏ Patient Experience Manager	Pilot a digital trust thermometer embedded in patient portals; stratify results by language and income, and map friction points by region.
❏ Community Health Worker Supervisor	Host monthly "digital mistrust listening hours" in under-connected neighborhoods; collect stories and translate into dashboard flags or redesign sprints.
❏ Director of Snacks and Morale	Verify N95 fit; also verify fit between mission, metrics, and muffins.

The Zero-Distance Future

Imagine waking up in a world where your ZIP code no longer predicts your lifespan. A world where a woman in rural Montana can access a robotic surgeon with the same ease as a man in downtown Mumbai. Where an Afghan refugee in Berlin can review her digital health record in her own language; where an unhoused veteran in Syracuse sees a street medicine doctor who adjusts his meds curbside; where a patient recovering from opioid use in Chicago is never cut off by an insurance clock that doesn't understand what healing really takes. In this world, a single mother in Appalachia no longer has to choose between a day's wages and a check-up. In this world, care shows up, wherever we are.

That is not science fiction, but the logical end of what we've already begun. We have the tools. What we need now is the will.

This chapter brings together the central themes of the book and looks squarely at what comes next. Built on the voices of patients and providers, engineers and ethicists, it explores what it will take to create a truly borderless healthcare system. Not a perfect one, but one where the **Physical, Financial, Digital, Cultural,** and **Trust/ Knowledge Barriers** shrink for all of us. From curbside wound care to tele-neurology in rural Virginia, from AI triage to street-level harm reduction, we are closer than ever to making access the rule rather than the exception.

But futures do not arrive on their own. They are built.

The Next Frontier: Emerging Technologies Changing the Game

If you ask a patient in rural Virginia, a street medic in upstate New York, a startup founder in California, or a policymaker in Kigali to describe the future of healthcare, you'll get very different answers. All of them will be right. Because the future isn't one thing: it's a field of tensions. Between what is possible and what is permitted. Between invention and implementation. Between whom a system is designed for and whom it leaves behind.

Across this book, you've seen how care is being redefined at the margins. In Chapter 5, Dennis Fowler reminded us that no surgical revolution—including laparoscopy, robotics, and telesurgery—ever begins in comfort. It begins in resistance. Yet the promise of remote skill-sharing is already emerging in places where specialists cannot go, and where mentorship must travel without a plane ticket. The real breakthrough is not in bandwidth, but in proximity to need.

In Chapter 9, we met Lucky, a transgender patient in Rochester who described a "dreamlike" care team built around them; not in spite of their complexity, but because of it. With a neurodivergent brain and a history of medical trauma, they found healing not in a new technology, but in a team that used tools intentionally. Their vision of decentralized, patient-controlled records isn't science fiction, it's survival strategy. It asks: What if control over your own data was not a feature, but a right?

These stories, and others like them, point to a central truth: the next frontier in healthcare is not a product. It's a posture. It asks not just what a tool can do, but who it is for. It considers not only what can be built, but what must be rebuilt: policy, infrastructure, reimbursement, and trust. And it insists that reach means nothing without relevance.

The genomics boom offers breathtaking breakthroughs, but access to that future fractures along painfully familiar lines. In Chapter 8, we examined how even modest digital tools can become unaffordable when reimbursement fails or infrastructure lags. Genomics raises the stakes. One patient may benefit from precision oncology, paying out-of-pocket for advanced sequencing; another cannot access basic genetic screening. One rides the wave, the other never even sees the water. Without intentional design and sustainable funding, innovation becomes a wedge that widens divides, not a bridge that brings care closer.

So yes, the future is here; in pieces, in patches, in pilot programs and scrappy startups and progressive ministries. But the next frontier is not about where AI is working or where robotics can go. It's about who gets to imagine, design, and benefit from those tools.

This is not a chapter about what's coming.

It's a chapter about what we're finally ready to build.

Breaking the Last Barriers: Policy and Practice in the Digital Age

Stories from the Field: Trust/Knowledge and Culture Barrier Spotlight[1]

Dr. Jewel Mullen knows what it means to carry the weight of a system on your shoulders and still choose to fight from within. A physician, former principal deputy assistant secretary for health at HHS, and now Associate Dean for Health Equity at Dell Medical School, she has spent decades navigating the intersection of practice, policy, and public health. But when you ask her what the future of healthcare access really demands, she cuts straight to the core: "The real change is going to have to come from undoing the capitalist healthcare system itself, and that's going to take people, including physicians, who are willing to put their professional lives on the line to bring about change. Inaction amounts to complicity. There is substantial documentation of the moral injury that results."

Dr. Mullen does not talk about disruption for its own sake. She talks about design; what it's designed to do, and who it's designed for. "We are seeing tech platforms being built not around need, but around market potential. Some people are devising solutions and then going out looking for a problem to apply them to," she said. "If the goal is cost control, not care, then the solutions will follow the money, not the person."

She warns that healthcare innovation is repeating history in fast-forward. "The future looks like the past," she told me. "It's the Wild West again. Tech outpacing policy, systems being built with no guardrails, and access being assumed rather than engineered." To shift the trajectory, Dr. Mullen calls for deeply intentional policy

---

[1] Mullen J. Interview with Matt S. Digital health, leadership, and healthcare transformation. Virtual interview. June 2025.

work. "Technology will always move faster than regulation, but that's no excuse to skip the work of asking, 'Who is this serving? What systemic inequity is this addressing?'" She points to the U.S. healthcare model's fundamental flaws, not just in coverage or cost, but in design logic. "Even if we got Medicare for All, the capitalist incentives embedded in the system would still be there. We're serving the wrong master."

She spoke of Taiwan's national system as a model of technology-leveraged inclusion, contrasting it with the American approach where "broadband is spotty, trust is fragmented, and tools are built for those already ahead." Her concern? That AI will not just reinforce disparities but encode them. "Will super-intelligent systems say, 'We only work in San Jose?' Will they determine rural populations are unprofitable and simply opt out?"

Mullen believes leadership, not tech, is the most urgent upgrade. "We need people who will stay in government and shape policy. We need people who will fight pharma's uncontrolled prices. And we need to stop designing approaches that prioritize serving individuals and families with comprehensive healthcare coverage while perpetuating two-tiered care systems."

But she is not advocating for perfection or ideology. She is advocating for honesty. "We know these innovations won't serve everyone. But we don't even design them to try." Her call is for policies that make healthcare work like infrastructure, not like luxury goods. At the end of our conversation, I asked her what kind of leaders we need in the decade ahead. Her answer was clear: "People who don't sell out. People who don't lose their vision just because the process is hard. People who remember that healthcare is about humanity."

The Promise: A World Where Geography Never Limits Care

Policy doesn't live on paper; it plays out in people. And while leaders like Dr. Mullen call out the systemic flaws that need reimagining, others are working within those systems to redesign the very levers we often overlook. From Medicare reimbursement to staffing structures, access isn't just about digital reach; it's about rebuilding the spine of healthcare itself. Dr. Sarah Corley knows that better than most.

Stories from the Field: Cultural and Digital Barriers Spotlight[2]

Dr. Sarah Corley has shaped the digital future of American healthcare; but for me, she also shaped my own. She was my first boss in tech. For the first month we worked together, I jumped every time I heard her name. (Seriously, jumped right out of my skin!) I was in way over my head, and she threw me in the deep end. And that has made all the difference.

A physician by training, now Chief Medical Advisor at the MITRE corporation, Sarah has built her career on pragmatic system redesign. She works at the intersection of clinical care and national infrastructure, advising agencies like the VA on what truly enables access and what just clutters the conversation.

"We need to augment our workforce by redefining who delivers care," she said. "In other countries, we use community health workers. We should be doing that here, especially in places like inner cities and rural America." But she's clear-eyed about the resistance. "You're going to face turf battles from doctors' and nurses' organizations, but the shortage is too big to ignore."

She also challenged the cultural lens around primary care. "It means different things in different places. And in our culture, which is fine popping the next pill and burning out, prevention has become taboo." As a physician, she finds that deeply troubling. "The best way to treat something is to keep from getting it in the first place. And we know that a relationship with a primary care provider lowers costs and improves outcomes."

Her proposal is straightforward and radical: "Make medical school free for people who enter and stay in primary care, not just a four-year commitment. And make the job less miserable." She believes ambient AI technology is already helping. "It's not about time saved. It's about emotional relief. Being able to look your patient in the eye again instead of a screen? That changes the whole experience of care."

Inbox overload, she added, is an underrecognized barrier to access. "There are too many messages," she said. "Some systems use AI to route messages or auto-reply when people send too many in a row. That's smart triage. But saying 'we'll have AI answer them all'? That's not the solution." Instead, Corley recommends a root cause analysis. "Maybe we're not spending enough time with our patients in person.

---

[2] Corley S. Interview with Matt S. Digital health, leadership, and healthcare transformation. Virtual interview. June 2025.

Maybe that's why the messages pile up." She's also wary of unvalidated AI chart summaries. "Sometimes it documents things you didn't do, because it assumes what usually happens. That's a risk. The control? Always link to the source note so you can verify what it generated."

Her vision of geographically unrestricted access depends on structure, not slogans. "When I started out, nursing staff did a lot more. Now they're all stuck in EHRs. Medicare reimbursements keep going down, and inflation keeps going up. You can't run a system like that and expect people to stay."

In Corley's world, the future of access looks less like disruption and more like design. "Train for the outcomes you want. Build systems that respect everyone's time. And fund what actually improves care. If we want access without borders, we have to build it from the inside out."

Making It Happen: Steps Toward Truly Borderless Healthcare

If design is the scaffolding of change, leadership is the load-bearing wall. Systems fail or flourish not just because of what they build, but because of who's allowed to lead. For every promising tool or scalable platform, there is a governance structure that either propels it forward or strangles it quietly. And few have seen that dynamic more clearly than Ed Marx.

Stories from the Field: Digital and Cultural Barriers Spotlight[3]

Ed Marx has built a career at the intersection of technology, clinical systems, and bold leadership. A former CIO for Cleveland Clinic, NYC Health + Hospitals, and Texas Health Resources, he's led digital transformation for some of the largest healthcare systems in the country. Today, as CEO of Marx Advisory and a widely respected strategist and board member, he consults across the globe on how to turn tech potential into health system reality. But ask him why transformation fails, and his answer has little to do with technology.

"It's not a change-management problem," Ed told me. "It's a governance problem. The leadership of these organizations is not progressive.

---

[3] Marx E. Interview with Matt S. Digital health, leadership, and healthcare transformation. Virtual interview. June 2025. Accessed July 29, 2025.

Therefore, they are unable to understand progressive ideas or make progressive decisions."

In Cleveland, he encountered firsthand how some decisions were being shaped by people not always equipped to navigate digital change. "They were very paper based, very old-fashioned. And yet they were the ones preventing me from doing a transcontinental mitral valve heart replacement. How can you lead digital transformation when you are not digitally transformed yourself?"

Even more revealing was what he experienced in Texas. "I got pulled aside by the head of supply chain," he said. "He had come from the partner vertical and told me, 'I came in just like you; gung-ho, ready to change the world. But I quickly realized this place doesn't want to move that fast. Dallas is a nice place to raise a family, the pay's good, the benefits are great. Eventually, I just toned it down.'"

This quiet normalization of stagnation, Ed explained, is what derails even the best intentions. "You come in ready to change the world, but the organization isn't. So, you have three choices: get fired, be miserable, or assimilate. And most people assimilate." For Ed, the antidote is not just bold vision but bold structure. "You've got to be a strong communicator," he said. "You need to understand politics, cast a clear vision, and surround yourself with the right people (people who will watch your left, your right, and your back) because there will always be someone trying to destroy what you're building."

One of the few leaders he saw navigate this well was Dr. Toby Cosgrove. "Toby was bold. He kicked McDonald's off campus. He banned employee smoking. And he said, 'If you can't close your lab coat, maybe you shouldn't work here.' But he also had charisma. And he built a team around him that protected the mission. That's why he thrived at Cleveland Clinic for 15 years."

Ed is candid about his own track record. "I've had success in some places and failed in others. In Cleveland I was doing some pretty bold things, but I didn't yet understand the political terrain. I didn't have my flanks protected. That made it harder."

His message is clear: the greatest barrier to borderless care is not a lack of tools or talent, but cultures that punish initiative and reward complacency. And until governance becomes as agile as the technology it hopes to deploy, transformation will continue to lag where it matters most—on the ground, with patients.

These stories are not just about policy or practice: they're about power. Who shapes the tools? Who sets the rules? And most importantly, who refuses to settle for what's always been? If we want a borderless healthcare future, we need more than new technology.

We need courage, continuity, and a new kind of leadership that builds for the many, not just the few.

What We Imagine, We Must Now Make Real

This book has taken you across continents, across care settings, and across boundaries that were once thought immovable. You've met street clinicians dispensing wound care beside food trucks; patients in Rwanda receiving algorithm-supported cancer diagnoses; and rural surgeons exploring remote robotic mentorship as a way to bring skill where no specialist is in reach. Across every story, one truth emerged: *distance in healthcare isn't just measured in miles. It's measured in power, infrastructure, identity, and trust.*

The borderless healthcare revolution is already underway. But a revolution is only as strong as those willing to carry it forward. What we imagine, we must now make real.

This is not a call for more disruption for disruption's sake. It is a call for deeply intentional change (just like Dr. Mullen called for!)— for patients to be seen, systems to be redesigned, and technology to serve everyone, not just efficiency. Whether you are a patient navigating your next appointment, a clinician shaping daily workflow, or a policymaker deciding where funding flows, *you are not on the sidelines. You are part of this movement.*

If you are a patient:

- Ask if your care can be remote.
- Demand full access to your records and carry them with you.
- Use digital tools, but leave them behind without hesitation when they fail you.

If you are a provider:

- Make care fit patients, not paperwork.
- Offer telehealth like you mean it.
- Document in ways that affirm, not obscure, identity.

If you are a policymaker or builder:

- Center edge-case users from day one.
- Build systems that adapt to culture, language, and lived realities.
- Fund trusted community liaisons who bridge care with credibility.

Checklist for a zero-distance future:

- Are your systems physically reachable, financially navigable, and culturally safe?
- Are digital tools closing gaps or creating new ones?
- Are you building for speed or for trust?
- Are patients forced to adapt to systems or are systems reshaped to reflect the people they serve?

This is not the end of the story. It's the handoff.

The borderless healthcare revolution is not about flattening the world. It's about lifting the barriers that never should have existed in the first place.

Zero-distance care is not a fantasy. It's a framework. It is not the work of one innovator, one clinic, or one policy.

It is the collective work of all of us, baking something new.

The borders are already falling.

Now, we build what comes next.

⚒ Startup Builder's Box: What to Build Next!

For health tech startups, clinical entrepreneurs, and digital rebels— here's how you can really make care borderless!

BUILD THIS	NOT THAT
Mobile-first, data-light, offline-resilient UX	Desktop-heavy, broadband-hungry UI
Text and WhatsApp as defaults	Portal logins and multifactor authentication
Multilingual on-boarding	English-only intake forms
PIN, faceID, or passcode-less entry	Traditional password gates
Care pilots in real clinics	Demos at tech conferences
Design for digitally hesitant users	Assumption of digital-native fluency
AI transparency features	Blackbox automation
Bias audits by ZIP code, race, identity, disability	One-size-fits-all performance claims
Human backup when tech fails	App-only experiences with no fallback

(continues)

Startup Builder's Box: What to Build Next! (continued)

BUILD THIS	NOT THAT
"Trust" as a tracked KPI	Vanity engagement metrics
Patient-paid insights	Data mining without consent or value exchange
Trauma-aware UX design	Bureaucratic, accusatory workflows
Opt-out friendly flows	Sticky traps that assume retention is loyalty
Simplified scheduling and billing	Digitized versions of broken forms
Community-first partnerships	Top-down hospital-only pilots
Inclusive by default (gender, disability, language)	Add-on accessibility later (if ever)

Bottom line: Don't just digitize healthcare. *Humanize it. Localize it. Make it earn trust.*

Leadership Checklist: From Pillars to Practice

I've told the stories. Here's what to do. Whether you're a clinician, policymaker, technologist, or community leader, these 25 steps translate the five pillars (**Physical, Financial, Cultural, Digital, and Trust/ Knowledge**) into action. One for every year since the EHR was supposed to fix this.

1. **Design for Dignity** *(Physical, Cultural, Trust)*
   - Audit your intake forms. If gender options still say "M/F," you're excluding care.
   - Map your access deserts. Identify where geography, language, or stigma block care.
   - Design for trust. If people do not show up, ask what your space is signaling.
   - Make "quality" portable. Bring your best workflows to your worst-connected sites.

2. **Connect the Edges** *(Physical, Digital, Cultural)*
   - Build hub-and-spoke specialty networks. Start with robotics, derm, and psych.
   - Fund mobile, pop-up, and street-based care, not just more walls.
   - Use telehealth to build continuity, not just plug gaps.
   - Make texting the default. No one should need a portal login to ask for help.

3. **Train with Intention** *(Trust, Cultural, Digital)*
   - Make cultural humility and digital literacy mandatory. Not nice-to-haves.
   - Train clinicians to use and question AI. Both skills matter.
   - Value lived experience as a credential. Build teams that reflect that.
   - Pay community workers and students for their expertise, not just their time.

4. **Measure What Matters** *(Trust, Digital, Financial)*

- Stop measuring access by wait times. Ask who never made it through the door.
- Disaggregate your data by ZIP code, language, disability, and race.
- Track digital dropout rates. That is your invisible cliff.
- Measure trust. Survey for it. Fund responses to it.

5. **Fund the Infrastructure of Access** *(Financial, Digital, Cultural)*

- Budget for phones, broadband, and translation; these are clinical necessities.
- Offer small incentives like $25 gift cards to improve uptake. They work.
- Create flexible funds for what keeps people stable—food, rides, meds.
- Invest in upstream ROI. This includes mobile mental health, gender-affirming care, addiction street teams.

6. **Lead Like It Matters** *(Trust, Cultural, Financial)*

- Put lived experience in the boardroom, not just in advisory groups.
- Stop calling people "noncompliant" when systems were not built for them.
- Champion change even when it costs you. That's leadership.
- Ask. Listen. Then fund what people actually say they need.
- Do not be an %$#&*@!. Build systems your grandmother would trust.

A Word from the Director of Snacks and Morale

Hi. It's me. The Director of Snacks and Morale.

You may have seen me lurking near the fruit bowl, the muting button, or the bad coffee in conference rooms where strategy goes to die.

Let me be clear: I have no formal authority.

I have no grant. No credentials. No keynote slot.

But I've been here the whole time.

- Filling the silent gaps when patients don't trust the portal.
- Walking people to the elevator after someone said "noncompliant" one too many times.
- Reminding teams that digital doesn't mean distant, and access isn't an app.

So now that you've read the book and scribbled all over the margins with "OMG YES," I have just one request:

Don't wait for permission.

Build something better.

Start the program. Write the policy. Text the patient. Burn the forms.

Order the snacks.

This is not about perfect. It's about closer.

Closer to trust. Closer to inclusivity. Closer to people.

See you in the breakroom. I'll be the one with the good chips. #StayCrispy

—*Director of Snacks and Morale, Chief Vibes Officer, Co-Conspirator of Hope, Sister in Disruption*

Acknowledgments

To my family, especially Gus and our four boys—thank you for the space, the fire, and the fuel. You've kept me grounded while I built something bold.

Index

Note: Page numbers in *italics* indicate diagrams.

Let's improve access for all, together.

Visit **DrSarahMatt.com**
for ongoing support in removing barriers
to healthcare through technology:

-Articles
-Resources
-Media and Video
-Engagements
-and more

DrSarahMatt.com